"十三五"应用型人才培养规划教材

ASP.NET网络数据库

刘保顺 ◎ 编著

清华大学出版社
北京

内 容 简 介

本书介绍了在 ASP.NET 框架下采用 C♯访问 SQL Server 数据库的编程技术。主要内容包括 SQL Server 中常用的 SQL 语法、三层架构下利用 ADO.NET 的 DataAdapter、DataReader、DataSet、DataTable、Connection、Command 等对象访问 SQL Server 的过程；采用 LINQ 访问数据和数据库编程，涉及的 LINQ 技术有利用 LINQ to Object 查询内存中的集合和数据、利用 LINQ to XML 查询 XML 中的数据、利用 LINQ to Entities 操作数据库等；GridView、Chart、UpLoad、TreeView 4 种 ASP.NET 内部控件，以及 ASP.NET Web 服务从网上获取信息的技术。

除 C♯访问数据库的服务器端编程外，本书还介绍了 HTML、CSS、JavaScript、jQuery 和 jQuery EasyUI 等基础知识和客户端的编程，服务器端与客户端以 JSON 格式进行数据交换的过程，以及 jsTree 插件制作树、OrgChart 插件制作组织结构图的编程。

为避免枯燥，全书将众多网络编程的知识点融入教学示例中，并配有教学 PPT 和习题答案，以方便教学。

本书适合 ASP.NET 初学者及对 ASP.NET 感兴趣的技术人员阅读。

本书封面贴有清华大学出版社防伪标签，无标签者不得销售。
版权所有，侵权必究。侵权举报电话：010-62782989 13701121933

图书在版编目(CIP)数据

ASP.NET 网络数据库/刘保顺编著. —北京：清华大学出版社,2019
("十三五"应用型人才培养规划教材)
ISBN 978-7-302-52822-7

Ⅰ.①A… Ⅱ.①刘… Ⅲ.①网页制作工具－程序设计－高等学校－教材 Ⅳ.①TP393.092.2

中国版本图书馆 CIP 数据核字(2019)第 082672 号

责任编辑：王剑乔
封面设计：刘　键
责任校对：李　梅
责任印制：宋　林

出版发行：清华大学出版社
网　　址：http://www.tup.com.cn, http://www.wqbook.com
地　　址：北京清华大学学研大厦 A 座　　邮　编：100084
社 总 机：010-62770175　　邮　购：010-62786544
投稿与读者服务：010-62776969, c-service@tup.tsinghua.edu.cn
质量反馈：010-62772015, zhiliang@tup.tsinghua.edu.cn
课件下载：http://www.tup.com.cn,010-62770175-4278

印 装 者：三河市少明印务有限公司
经　　销：全国新华书店
开　　本：185mm×260mm　　印　张：20　　字　数：483 千字
版　　次：2019 年 6 月第 1 版　　印　次：2019 年 6 月第 1 次印刷
定　　价：59.00 元

产品编号：082226-01

前言
FOREWORD

　　网络编程涉及的知识点较多，既有后台程序的开发，也有前台界面的编程。ASP.NET后台开发的语言目前多以C#为主，前台界面开发有许多优秀的框架，而jQuery就是最受欢迎的框架之一。目前，市面上有许多关于ASP.NET的书籍，其多以介绍各种控件为主，但这一编程方式由于程序深度开发时受限，如很难与jQuery架构整合，正在逐渐退出市场。在当今大数据时代，数据交换的格式除XML外，更多的是JSON。

　　本书是一本介绍使用jQuery、C#访问和操作SQL Server数据库的教程。书中知识点的介绍融入教学示例中，围绕示例，循序渐进，由浅入深，尽量避免枯燥的知识性的讲解。

　　本书采用Visual Studio 2012作为程序开发平台，SQL Server作为数据库管理系统，讲述ASP.NET开发Web程序涉及的诸多知识点，包括HTML、CSS、JavaScript、jQuery、C#、SQL Server、ADO.NET、LINQ、Web Service等。本书只介绍少量的内部控件如GridView、UpLoad、Chart、TreeView外，其他内部控件介绍得很少。书中以大量的篇幅介绍ASP.NET程序开发中浏览器端的编程知识，如HTML、CSS、jQuery和jQuery EasyUI；介绍三层架构下服务器端使用ADO.NET对象模型、访问SQL Server数据库的方法；介绍浏览器端与服务器端以JSON格式进行数据交换的过程。

　　各章节安排如下。

　　第1章　ASP.NET开发和运行环境。讲述IIS的安装、ASP.NET网页程序的运行、ASP.NET程序开发的模式、本地网页上传到云服务器等。

　　第2章　数据库基础知识。讲述关系数据库管理系统中关系模型、表间关系、建立表的原则、SQL Server基本操作、SQL的主要语法。

　　第3章　网页编程基础知识。讲述HTML、CSS、JavaScript基本语法和对象；JavaScript解析JSON；XML和AJAX基本概念；浏览器和服务器通过JSON交换数据的过程。

　　第4章　ASP.NET内置对象。介绍Page对象的属性和事件、Response、Request、Session、Application和Server对象的属性与方法。

　　第5章　C#语言基础知识。介绍C#中类的定义、封装、继承和多态性；数组和泛型的使用、服务器端JSON的序列化和反序列化的方法。

　　第6章　ADO.NET连接和命令对象。介绍通过Connection连接SQL Server数据库的方法、使用Command和DataReader读取SQL Server表中的数据；基于三层架构，建立可重用的访问数据库的过程、注入式SQL攻击的防范、利用存储过程访问和操作数据库、ASP.NET应用程序中的事务。

第 7 章　非连接的数据访问对象和工厂模型。介绍 DataTable 中数据的读取、删除和更新，DataSet/DataTable 的数据转换为 XML，DataSet/DataTable 对象的序列化，DataSet 中建立表间的关系，建立与提供程序无关的程序代码。

第 8 章　ASP.NET 内部控件。利用 Chart 将数据库中数据以图形方式展示，利用 TreeView 控件将数据库中数据以树形显示，利用 UpLoad 控件实现文件的上传。

第 9 章　jQuery。jQuery 的基本语法、jQuery 的事件、jQuery 操作 HTML 和 CSS、jQuery 操作表单、使用 jQuery 以 AJAX 方式提交表单数据、浏览器解析 XML 和 JSON 格式的数据、jquery.cookie.js、jsTree 和 OrgChart 的使用。

第 10 章　语言集成查询——LINQ。介绍使用 LINQ to Object 查询内存中的集合和数据、LINQ to XML 查询 XML 中数据、LINQ to Entities 操作数据库。

第 11 章　ASP.NET Web 服务。介绍 ASP.NET Web 服务的建立和使用、使用 Newtonsoft.Json 创建提供 JSON 格式的 ASP.NET Web 服务。

第 12 章　jQuery EasyUI。jQuery EasyUI 使用的方法、页面的布局、消息框和 datagrid 的使用。

本教材已列入北京科技大学校级规划教材，教材的编写和出版得到了北京科技大学教材建设经费的资助，在此深表感谢。

编著者
2019.1

本书配套教学课件、教学示例及习题答案.rar

目 录
CONTENTS

第1章 ASP.NET 开发和运行环境 ·· 1
  1.1 Web 应用程序的工作原理 ··· 1
  1.2 静态网页和动态网页 ·· 2
  1.3 动态网页开发技术 ·· 4
  1.4 IIS 的安装和配置 ··· 5
  1.5 将本地站点上传到云 ·· 6
  1.6 ASP.NET 程序开发环境和开发模式 ································ 8
    1.6.1 Web 窗体拖曳控件的方式 ·· 8
    1.6.2 MVC 模式 ··· 11
    1.6.3 使用原生态 Web 表单模式 ····································· 11
    1.6.4 三层架构 ··· 16
  习题和思考 ··· 16

第2章 数据库基础知识 ·· 17
  2.1 数据库管理系统 ··· 17
  2.2 数据库 ·· 18
  2.3 关系数据库 ·· 18
  2.4 数据库系统 ·· 19
  2.5 数据库中表间的关系 ·· 20
  2.6 SQL Server 的基本操作 ··· 21
    2.6.1 建立数据库 students ··· 21
    2.6.2 建立数据库关系图 ··· 23
    2.6.3 在数据库中增加记录 ··· 25
    2.6.4 权限设置 ··· 25
  2.7 关系数据库标准语言——SQL ·· 26
    2.7.1 单表数据查询 ··· 26
    2.7.2 排序 ··· 28
    2.7.3 统计和分组 ··· 28
    2.7.4 多表数据查询 ··· 31
    2.7.5 增加、删除和更新记录 ··· 32

  2.7.6 操作表结构的 SQL ·············· 33
 习题与思考 ·············· 33

第 3 章 网页编程基础知识 ·············· 36
 3.1 HTML ·············· 36
  3.1.1 HTML 文档结构 ·············· 36
  3.1.2 HTML 表单 ·············· 38
  3.1.3 表格 ·············· 41
 3.2 CSS ·············· 42
 3.3 通过 JavaScript 为网页增加动作 ·············· 46
 3.4 用 JavaScript 修改 HTML 元素的样式 ·············· 49
 3.5 JavaScript 编写简单的扑克游戏 ·············· 50
  3.5.1 数组对象 ·············· 51
  3.5.2 自定义对象 ·············· 52
  3.5.3 扑克牌中的页面 ·············· 54
  3.5.4 扑克牌中的属性和方法 ·············· 54
 3.6 window 对象控制定时效果 ·············· 57
 3.7 XML ·············· 59
 3.8 JavaScript Object Note ·············· 60
  3.8.1 JSON 数据格式 ·············· 60
  3.8.2 JSON 文本串转换为 JavaScript 对象 ·············· 63
  3.8.3 将 JavaScript 对象转换为 JSON 字符串 ·············· 64
  3.8.4 ASP.NET 中浏览器和服务器通过 JSON 的数据交换过程 ·············· 64
 习题与思考 ·············· 67

第 4 章 ASP.NET 内置对象 ·············· 71
 4.1 Page 对象 ·············· 72
 4.2 Response 对象 ·············· 74
  4.2.1 Response 对象的属性和方法 ·············· 74
  4.2.2 Response 对象应用示例 ·············· 75
 4.3 Request 对象 ·············· 76
  4.3.1 Form 集合 ·············· 77
  4.3.2 QueryString 集合 ·············· 78
  4.3.3 ServerVariables 集合 ·············· 78
 4.4 Server ·············· 79
  4.4.1 Transfer() 方法 ·············· 79
  4.4.2 MapPath() 方法 ·············· 80
  4.4.3 HTML 和 URL 编码 ·············· 80
 4.5 ASP.NET 状态管理 ·············· 80
  4.5.1 ViewState 对象 ·············· 80
  4.5.2 Cookies ·············· 83

4.5.3　Session ……………………………………………………… 86
　　　4.5.4　Application …………………………………………………… 87
　习题与思考 ……………………………………………………………… 89

第5章　C#语言基础知识 …………………………………………… 92
　5.1　类 …………………………………………………………………… 92
　　　5.1.1　类的定义 …………………………………………………… 92
　　　5.1.2　使用类建立对象 …………………………………………… 94
　　　5.1.3　类的封装 …………………………………………………… 97
　　　5.1.4　类的继承 …………………………………………………… 100
　　　5.1.5　类的多态性 ………………………………………………… 101
　5.2　集合与泛型 ………………………………………………………… 103
　　　5.2.1　泛型集合 List<T> 的使用 ………………………………… 104
　　　5.2.2　泛型集合 Dictionary<Key,Value> 的使用 ……………… 105
　5.3　其他数据类型 ……………………………………………………… 106
　　　5.3.1　DateTime 和 TimeSpan …………………………………… 106
　　　5.3.2　Convert 类 ………………………………………………… 108
　　　5.3.3　String 类 …………………………………………………… 108
　　　5.3.4　System.Text.StringBuilder 类 …………………………… 109
　5.4　委托 ………………………………………………………………… 109
　5.5　JSON 的序列化和反序列 ………………………………………… 110
　　　5.5.1　使用 JavaScriptSerializer 类序列化和反序列化 ……… 111
　　　5.5.2　使用 formatter 格式化器序列化和反序列化 …………… 113
　　　5.5.3　使用 Json.NET 序列化和反序列化 ……………………… 114
　习题与思考 ……………………………………………………………… 115

第6章　ADO.NET 连接和命令对象 ……………………………… 119
　6.1　ADO.NET 基础 …………………………………………………… 119
　6.2　Connection 对象 ………………………………………………… 121
　6.3　Command 类 ……………………………………………………… 123
　6.4　DataReader 类 …………………………………………………… 125
　6.5　构建可重用的访问数据库的代码 ………………………………… 127
　　　6.5.1　数据访问层 ………………………………………………… 127
　　　6.5.2　建立连接的逻辑 …………………………………………… 130
　　　6.5.3　在 DBbase 类中建立查询数据的逻辑 …………………… 130
　　　6.5.4　在 DBbase 类中建立插入数据的逻辑 …………………… 132
　　　6.5.5　在 DBbase 类中建立更新数据的逻辑 …………………… 133
　　　6.5.6　界面层的设计 ……………………………………………… 133
　6.6　SQL 注入攻击 …………………………………………………… 134
　6.7　参数化命令 ………………………………………………………… 136
　6.8　存储过程 …………………………………………………………… 137

   6.8.1 建立 SQL Server 的存储过程 ……………………………… 137
   6.8.2 在 DBbase 类中建立查询数据的逻辑 …………………… 139
   6.8.3 在 DBbase 类中建立插入数据的逻辑 …………………… 140
   6.8.4 在 DBbase 类中建立更新数据的逻辑 …………………… 141
   6.8.5 数据输入界面 …………………………………………… 141
   6.8.6 显示数据页面 …………………………………………… 142
  6.9 事务 ……………………………………………………………… 143
   6.9.1 存储过程事务 …………………………………………… 143
   6.9.2 ADO.NET 事务 ………………………………………… 144
  习题与思考 ……………………………………………………………… 147
第 7 章 非连接的数据访问对象和工厂模型 ……………………………………… 148
  7.1 DataSet 类 ……………………………………………………… 149
  7.2 DataSet 类的主要属性和方法 ………………………………… 149
   7.2.1 DataSet 的主要属性 …………………………………… 149
   7.2.2 DataSet 的主要方法 …………………………………… 150
  7.3 DataTable 类 …………………………………………………… 152
   7.3.1 DataTable 的使用 ……………………………………… 152
   7.3.2 获取 DataTable 中的数据 …………………………… 153
   7.3.3 DataTable 中删除和更新记录 ……………………… 155
  7.4 DataAdapter 类 ………………………………………………… 157
   7.4.1 使用 DataAdapter 填充 DataSet ……………………… 158
   7.4.2 将 DataSet/DataTable 对象序列化为 XML …………… 159
   7.4.3 将 DataSet/DataTable 对象以二进制格式序列化 …… 160
   7.4.4 将 DataSet/DataTable 对象序列化为 JSON …………… 161
   7.4.5 DataSet 充填多个表和关系 …………………………… 164
   7.4.6 利用 DataAdapter 更新数据库中的数据 ……………… 166
  7.5 DataView 类 …………………………………………………… 167
  7.6 在数据访问类中使用 DataSet ………………………………… 168
  7.7 编写与提供程序无关的程序代码 ……………………………… 168
   7.7.1 创建工厂 ………………………………………………… 169
   7.7.2 使用工厂建立对象 ……………………………………… 169
   7.7.3 使用与程序无关的代码查询示例 ……………………… 170
  习题与思考 ……………………………………………………………… 172
第 8 章 ASP.NET 内部控件 …………………………………………………… 175
  8.1 GridView 控件 ………………………………………………… 176
  8.2 Upload 控件 …………………………………………………… 180
  8.3 Chart 控件 ……………………………………………………… 181
   8.3.1 Chart 控件添加数据 …………………………………… 183
   8.3.2 Chart 控件数据绑定 …………………………………… 183

8.3.3 制作数据回归曲线 188
8.4 TreeView 控件 194
习题与思考 198

第 9 章 jQuery 200

9.1 jQuery 概述 200
 9.1.1 jQuery 的作用 200
 9.1.2 下载和引用 jQuery 201
 9.1.3 用 jQuery 处理 DOM 201
 9.1.4 显示和隐藏小狗的示例 201
9.2 jQuery 选择器 203
9.3 jQuery 代码执行的时机和事件 206
9.4 jQuery 动态效果的函数 207
 9.4.1 显示和隐藏 207
 9.4.2 滑动函数 207
 9.4.3 淡入淡出函数 208
9.5 jQuery 对 HTML/CSS 操作 208
9.6 jQuery 操作表单 210
 9.6.1 表单中元素的选择器 210
 9.6.2 jQuery 操作表单中的元素 210
9.7 jQuery 与 Ajax 213
 9.7.1 Ajax 方式提交数据 214
 9.7.2 浏览器解析 XML 数据 228
9.8 使用 jquery.cookie.js 230
9.9 使用 jsTree 制作 tree 232
9.10 使用 OrgChart 制作组织结构图 236
习题与思考 244

第 10 章 语言集成查询——LINQ 248

10.1 LINQ to Objects 249
 10.1.1 LINQ 查询语法和步骤 249
 10.1.2 LINQ 查询表达式 251
 10.1.3 LINQ 的立即执行 254
 10.1.4 方法查询 254
10.2 LINQ to XML 255
10.3 LINQ to Entities 258
 10.3.1 生成数据模型 259
 10.3.2 LINQ to Entities 查询 260
 10.3.3 LINQ to Entities 数据库操作 262
习题与思考 262

第 11 章　ASP.NET Web 服务 …… 264
11.1　Web 服务的应用 …… 264
11.2　创建提供查询学生成绩的 Web 服务 …… 270
习题与思考 …… 277

第 12 章　jQuery EasyUI …… 278
12.1　jQuery EasyUI 概述 …… 278
12.2　jQuery EasyUI 的 Layout …… 279
12.3　对话框 …… 281
12.3.1　$.messager.show(options) …… 281
12.3.2　$.messager.alert(title,msg,icon,fn) …… 282
12.3.3　$.messager.confirm(title,msg,fn) …… 283
12.3.4　$.messager.prompt(title,msg,fn) …… 283
12.3.5　$.messager.progress(options or method) …… 284
12.4　form …… 284
12.4.1　form 提交数据 …… 284
12.4.2　form 加载数据 …… 288
12.5　jQuery EasyUI 的 Datagrid …… 290
习题与思考 …… 305

参考文献 …… 308

ASP.NET开发和运行环境

（1）静态网页和动态网页。
（2）ASP.NET 开发环境。
（3）ASP.NET 程序开发的方式。
（4）ASP.NET 程序的运行。
（5）本地网页上传到云。

1.1 Web 应用程序的工作原理

1. 服务器端和客户端

Web 是基于客户端/服务器的一种体系结构。通常将提供服务的一方称为服务器，接受服务的一方称为客户端。如当用户浏览某门户网站时，该门户网站就是服务器，用户使用的计算机或者手机等智能设备就是客户端。只要在计算机上安装有提供服务的软件，这台计算机就变成一台服务器，如在计算机上装上 SQL Server 数据库管理系统，该计算机就是一台数据库服务器。Web 服务器（或 WWW 服务器）指的是装有能够接受和响应来自客户端计算机请求的特定软件的计算机，如 Apache 服务器、IIS（Internet Information Server）服务器等。在一台计算机上装上客户端软件，该计算机就成为客户端。在因特网中客户端软件一般是指浏览器软件，如微软的 IE、Mozilla Firefox、Google 的 Chrome 等。

服务器与客户端的通信过程整体可以分为以下几个过程，如图 1-1 所示。

（1）用户在客户端浏览器输入要请求的 URL，并按 Enter 键发送这个请求。

（2）服务器根据请求的 URL 判断客户端的请求是静态网页还是动态网页。如果请求的是静态网页，服务器找到该网页，原样送回到浏览器；如

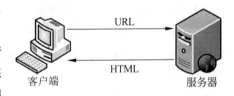

图 1-1 服务器与客户端通信过程示意图

果请求的是动态网页,服务器就会编译、执行用户请求的文件,生成标准的 HTML 文件,然后将这个 HTML 送回到客户端。

(3) 静态网页的文件由客户端上的浏览器负责解释,将解释后的结果显示在用户浏览器上。

2. 超文本传送协议(HTTP)

HTTP(HyperText Transfer Protocol)是一种以 TCP/IP 通信协议为基础的应用协议,它提供了 WWW 服务器和客户端浏览器之间传递信息的一种机制。HTTP 会话包括以下几个过程。

(1) 客户端与服务器建立连接。
(2) 客户端向服务器发出请求。
(3) 如果请求被接受,则服务器将响应结果送回至客户端。
(4) 客户端与服务器断开连接。

3. URL

因特网上每个网页都具有一个唯一的名称标识,通常称为 URL(Uniform Resource Locator,统一资源定位符)地址,这种地址可以是本地磁盘、局域网上某台计算机,也可以是因特网上某个网站,其基本格式如下。

protocol://hostname[:port]/path/

protocol:指定的传送协议,通常为 HTTP 或 HTTPS。

hostname:存放提供服务资源的主机名或者 IP 地址。

port:端口号,省略时取默认值。各种传送协议都有默认的端口号,HTTP 默认端口号是 80。

path(路径):表示主机上的一个目录或者文件名,如果省略文件名,对 IIS 服务器会查找 index.html、index.htm、index.aspx、default.html、default.html、default.aspx(查找顺序可在 IIS 中指定)。

4. HTML

HTML(HyperText Markup Language)是超文本标记语言,标准通用标记语言下的一个应用。"超文本"就是指页面内可以包含图片、链接,甚至音乐、程序等非文字元素,其语言结构分为 head 和 body 两部分。HTML 控制文字、图片等在浏览器显示的大小、格式等,浏览器能够解释 HTML。

1.2 静态网页和动态网页

1. 静态网页

静态网页主要是由 HTML 构成。虽然有的静态网页中包含有用 JavaScript 或 VBScript 编写的程序代码(一般称为脚本),但这些脚本(Script)是随着 HTML 一起从服务器传送到客户端的,其运行是在客户端上运行的,这种脚本语言称为客户端脚本。静态网页一经制成,内容一般不会发生改变,如果要修改其内容,一般需要修改源代码,然后重新上传到服务器。下面是一段用 HTML 编写的网页代码：

```
<html>
<head>
<title>my first web</title>
<style>
.headerH1{font-size:15pt;color:rgb(255,0,0)}
</style>
<script language="javascript">
function sayHello()
{
    alert('你好')
}
</script>
</head>
<body>
<span class="headerH1">ASP.NET 网络数据库</span>
<p onclick="sayHello()">click me</p>
</body>
</html>
```

在记事本中书写这些 HTML，保存为扩展名是.htm 或者.html 的静态网页。双击该文件，在 Google Chrome 浏览器中打开后效果如图 1-2 所示。这些代码包含了普通网页 HTML 的主要结构。在＜head＞标记中放置了＜title＞标记，定义网页标题，＜style＞标记定义了 CSS(Cascading Style Sheets，层叠样式表)控制"ASP.NET 网络数据库"文字的大小和颜色，＜script＞标记标识一段 JavaScript 脚本代码。

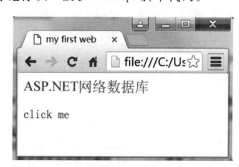

图 1-2　用 Chrome 浏览普通网页的结果

静态网页是保存在服务器上的文件，文件间基本上是独立存在的。静态网页没有数据库支持，内容相对稳定，Web 服务器查找方便，访问效率较高。静态网页工作过程如下。

（1）用户通过客户端的浏览器输入要访问的网页(扩展名一般是.htm 或.html)，发出 WWW 请求。

（2）服务器收到静态网页的请求。

（3）服务器从硬盘上指定的位置找到相应的 HTML 文件，发送给客户端。

（4）客户端的浏览器收到请求的文件，并且解释这些 HTML 代码，显示出来。

需要注意的是，静态网页中并非全是 HTML，在静态网页中也可以有 JavaScript 或者

VBScript 编写的程序代码,如上面的 HTML 代码中＜script＞标记中的代码,当单击 click me 时会弹出一个确认框。静态网页中的代码不是在服务器上运行,而是当客户端发出 URL 请求时,服务器将代码随着 HTML 一起发送到客户端,然后在客户端运行。由于这种网页的行为是在客户端进行的,是在客户端的"动",而网页本身是静态的,故这类网页仍然是静态网页。静态网页中的程序代码可通过 IE 浏览器中"查看"→"源"菜单命令查看。

2. 动态网页

动态网页是执行时用户可以根据输入所允许的各种信息,实现人机交互,能够根据不同的时间和地点、不同的访问者显示不同的内容,显示的内容一般来自数据库。动态网页中不仅包含 HTML 标记,还有各种相关的程序代码。动态网页工作过程如下。

（1）用户通过客户端的浏览器输入要访问的网址,发出 WWW 请求。

（2）服务器收到动态网页的请求。

（3）服务器从硬盘上指定的位置找到相应的文件,并且运行这些动态网页生成标准的 HTML。

（4）服务器将生成的 HTML 代码返回给客户端。

（5）客户端浏览器收到这些 HTML,解释这些 HTML 代码,并显示在计算机屏幕上。

与静态网页相比,动态网页的服务器不仅要找到这些网页,而且还要在服务器上运行这些文件,生成 HTML。这样动态网页是不能在客户端通过"查看"→"源"菜单命令查看到"源代码"的,看到的是服务器运行源代码后的结果。

1.3 动态网页开发技术

动态网页是在静态网页的基础上发展起来的。第一个真正使服务器能够根据执行时的具体情况生成动态 HTML 页面的技术是 CGI(Common Gateway Interface,通用网关接口)技术,文件的扩展名为.cgi。

1994 年出现了 PHP(PHP Hypertext Preprocessor,超文本预处理程序)语言,它将 HTML 代码和 PHP 指令合成为完整的服务器端动态页面,文件扩展名为.php。

1996 年微软借鉴 PHP 思想,在其 Web 服务器 IIS 3.0 中引入了 ASP(Active Server Pages,活跃服务器页面),迅速成为 Windows 系统下 Web 服务器端的主流开发技术,文件扩展名为.asp。

1997 年 Servlet 技术问世,1998 年 JSP(Java Server Pages)技术诞生。Servlet 和 JSP 的组合让 Java 开发者同时具有了类似 CGI 程序集中处理功能和类似 PHP 的 HTML 嵌入功能,文件扩展名都为.jsp。

2000 年微软推出了基于.NET Framework 的 ASP.NET 1.0 版本,以后新的版本不断推出,ASP.NET 可在服务器上生成功能强大的 Web 应用程序,文件扩展名为.aspx。

由于 Visual Studio 自带服务器环境,不安装 IIS 也可以在 Visual Studio 中调试 ASP.NET 程序,但如果要为客户端提供服务,如在校园网或者企业内部网上访问服务器上的网页程序,服务器上就必须安装并配置 IIS。

1.4 IIS 的安装和配置

1. 安装 IIS

ASP.NET 编写的动态网页,服务器端需要安装 IIS。Windows 默认情况下是不安装 IIS 的,如果要安装 IIS,需要 Windows 操作系统是专业版或者旗舰版。安装方法(以 Windows 7 为例):控制面板→程序→打开或关闭 Windows 功能,如图 1-3 所示。

图 1-3 IIS 的安装

在图 1-3 中,打开"Internet 信息服务",选中"Web 管理工具"复选框,以便兼容 IIS 6。在"万维网服务"中打开"常见 HTTP 功能",选中"静态内容"复选框。打开"应用程序开发功能",选中 ASP.NET 复选框,如果要兼容 ASP,选中 ASP。

2. 配置 IIS

右击桌面上的"计算机",在弹出的快捷菜单中选择"管理"→"服务和应用程序"命令,在弹出的对话框中选择"Internet 信息服务"→"连接"→"网站"命令,右击 Default Web Site,在弹出的快捷菜单中选择"添加应用程序"命令,弹出如图 1-4 所示对话框。

图 1-4 中选项说明如下。

"别名":输入一个有意义且好记的名字,这个"别名"在运行程序时要使用,这里输入的是 example。

"物理路径":存放网站应用程序的文件夹,可以在计算机的某盘上建立一个文件夹,将网站上的内容都放在该文件夹下。

3. 在本地计算机上运行网页程序

动态网页的程序不能双击运行。如果编程环境采用的是 Visual Studio(VS),大多数程序可以在 VS 环境中直接运行,但有些程序只能在浏览器中以 IIS 的方式运行:打开浏览

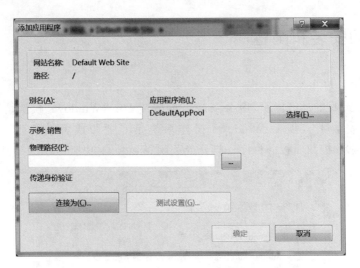

图 1-4　添加应用程序

器,在地址栏中输入"localhost:端口号/别名/网页名称"即可。

默认情况下的端口号是 80,可以省略不写,本书调试时由于 80 端口号已经分配给 PHP 程序,分配给 ASP.NET 网站的端口号是 81,如果要运行网站根文件夹下的 webform1.aspx 程序,在地址栏中输入 localhost:81/example/webform1.aspx 即可。

1.5　将本地站点上传到云

应用程序在当地计算机上调试完毕,需要发布到网上,让其他人通过 Internet 可以访问你的应用程序,最简单的方法是租用云服务。下面以租用阿里云为例,说明网站上传的过程。

登录 https://www.aliyun.com/网站,网上提供的云服务器 ECS(Elastic Computer Service)是一种可伸缩性的计算服务,根据个人需要,上面有不同的产品和价格,本书租用的是 1 核 2G,1M 固定带宽,40G 高速云盘。租用 ECS 后,相当于租用了一个远程虚拟的计算机。在租 ECS 时,可以选择操作系统,对 ASP.NET 而言,需要选择 Windows Server 作为服务器。阿里云提供了公网 IP(如 47.104.91.99)和私有 IP(如 172.31.70.48)。公网 IP 用于外部访问时使用,如通过 Internet 访问数据库,连接数据库的字符串的 IP 应该使用公网 IP。如果将网页传到云上,在云上虚拟的计算机上调试网页程序,或者登录云上的 SQL Server 数据库,需要使用私有 IP。用户还需要在云上配置 ASP.NET 的环境,包括 SQL Server 数据库、IIS 等。云上有镜像市场,上面有相应的环境。在购买 ECS 后,一般可以免费得到 ASP.NET 的运行环境。当地计算机登录云的操作步骤如下。

❶ 在计算机"开始"→"搜索程序和文件"的文本框中输入命令:mstsc,按 Enter 键后如图 1-5 所示。

❷ 在图 1-5 中输入远程计算机的 IP 地址,即阿里云分配的公网 IP,单击"选项"按钮,弹出"本地资源"对话框,单击"详细设置"按钮,弹出图 1-6 所示的"本地设备和资源"的设置对话框。如果要上传的网站在当地计算机的 D 盘,就选中 D 盘,这样当进入云计算机后,就在云上看到当地计算机 D 盘上的内容,可以直接将 D 盘上要发布的网站的内容方便地复制到云盘上。

图 1-5 连接远程计算机

图 1-6 "本地设备和资源"的配置

❸ 图1-6设置完成后,单击"确定"按钮返回后,再单击"连接"按钮,输入登录的密码(在申请云时设置的),即可登录到云计算机。系统已经自动配置好IIS,安装了SQL Server。双击SQL Server,登录到界面中,服务器可以输入虚拟机的名称,也可以输入公网IP(或者私有IP),登录到云上的SQL Server。

❹ 在云计算机上,单击"计算机",可以看到云盘C和当地计算机的盘符"D:",将D盘中的指定文件复制到C盘,按照配置当地计算机上IIS的方法配置云上的IIS。需要注意的是网页程序中连接数据库字符串为:

"server = 47.104.91.99;database = 数据库名;UID = 用户名;pwd = 密码";

如果直接在云计算机上调试网页程序,连接数据库和字符串中的IP地址可以是私有IP,也可以公网IP;如果要通过Internet访问数据库,就要将字符串中的IP地址改为公网IP。

说明:通过C♯访问阿里云上的MySQL,需要在云服务器管理控制台下的"云服务器ECS"→"网络和安全"→"安全组"下,配置运行实例的"安全规则",在"入方向"增加一条新规则,"协议类型"选择MySQL(3306),"授权对象"填写为0.0.0.0/0,否则不能访问阿里云的数据库。

1.6　ASP.NET 程序开发环境和开发模式

ASP.NET 开发环境最常用的是 Visual Studio。Visual Studio(VS)是微软推出的配合.NET 的 IDE 开发环境,其本身包含有.NET Framework 和 ASP.NET 程序开发的服务器。在编写程序时,其能够通过智能代码提示,给出目前可用的属性、方法和参数。通过该环境可以开发基于 Windows 的应用程序,也可以开发 ASP.NET 网页程序。其支持的语言有 Visual Basic、C♯、C++,本书使用的是 C♯。

打开 Visual Studio 软件,在"新建项目"对话框中选择 Web→"ASP.NET 空 Web 应用程序"选项,在"名称"文本框中输入 aspExample,在"位置"下拉列表框中选择 D:\aspExample,选中"为解决方案创建目录"复选框,如图 1-7 所示。单击"确定"按钮后,Visual Studio 自动在 D 盘上建立的文件夹为 D:\aspExample\aspExample\aspExample,在 D:\aspExample\aspExample 下生成两个解决方案的文件 aspExample.sln 和 aspExample.v11.suo。在以后的操作中单击 aspExample.sln 就可进入当前项目。网页代码放在 D:\aspExample\aspExample\aspExample 中。选中项目 aspExample 后右击"添加",在弹出的快捷菜单中选择"新建文件夹"命令,建立本书各章节的文件夹 chapter1、chapter2……各章节的示例放在相应的章节中。

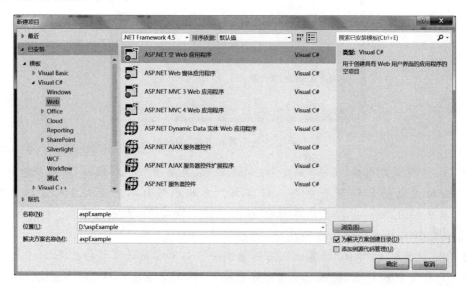

图 1-7　"新建项目"对话框

ASP.NET 网络程序开发,有以下几种模式。

1.6.1　Web 窗体拖曳控件的方式

开发人员在 Visual Studio 提供的可视化设计器中拖曳控件,然后编写代码响应事件。在此种模式下,开发 B/S 程序与 C/S 一样,这大大降低了程序开发的难度。但由于此种方式下控件中封装了许多内容,开发者很难了解背后的 HTML 是如何运行的,CSS、JavaScript、jQuery、Ajax 控制页面相对不灵活,程序不易于后期维护,不方便深度开发。目前窗体控件模式的开发正在退出微软.NET 程序开发的舞台。

例 1-1 在 Visual Studio 中使用窗体拖曳控件的方式，设计完成图 1-8 所示的表单，程序运行后，输出填写的内容。

❶ 在"解决方案资源管理器"中，右击 chapter1，在弹出的快捷菜单中选择"添加"→"Web 窗体"命令，在弹出的对话框中的"指定项名称"中输入 1_1.aspx，单击"确定"按钮。

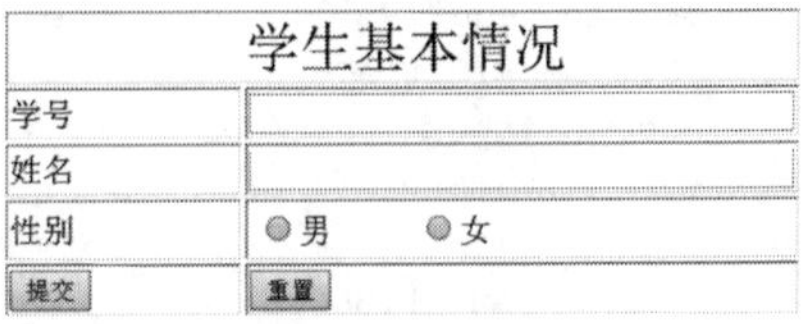

说明：VS 将同时生成 1_1.aspx、1_1.aspx.cs、1_1.aspx.desigers.cs 这 3 个文件。1_1.aspx 存

图 1-8 Web 窗体拖曳控件方式设计的表单

储的是页面设计时各个控件的代码，处理代码一般放在.cs 文件中；1_1.aspx.cs 是采用代码隐藏页模型设计网页时的代码隐藏文件，存储的是程序代码，一般存放与数据库连接和数据库相关的查询、更新、删除等操作代码，还有各个按钮单击后发生的动作代码等；1_1.aspx.desigers.cs 通常存放的是一些页面控件中的配置信息，就是注册控件页面，是窗体设计器生成的代码文件，作用是对窗体上的控件初始化。

❷ 打开"工具箱"，选中 HTML 下的 Table，拖放到 1_1 页面。默认情况下 table 显示为 3 行 3 列。

说明：在 VS 中网页有"设计""拆分""源"3 种视图。"设计"：以所见即所得的方式显示网页各元素的布局；"拆分"：同时显示布局和代码；"源"：显示 HTML 代码。"设计"视图下网页中元素改变时，"源"视图下的代码会自动地改变；"源"视图下直接改变代码，"设计"视图下网页的元素也会自动改变。

在网页设计中，可用于页面布局的有标记＜table＞和 DIV＋CSS，本示例使用的是标记＜table＞。

❸ 在 1_1.aspx 的"设计"视图下，通过菜单中的"表"→"修改"→"合并单元格"命令，再执行"表"→"插入"命令，增加行，使表变为 5 行 2 列（第 1 行列合并）。

❹ 将标记、文本框、RadioButtonList、命令按钮等分别放在表格的各单元格中，设置"学号""姓名"对应的文本框的 ID 分别为 txtStudent_no、txtStudent_name；命令按钮的 Text 属性分别设置为"提交""重置"；"性别"对应选项的 ID 属性设置为 RdoGender，单击 RdoGender 后面的设置属性按钮，将"性别"的 itsms 设置为图 1-9 所示。

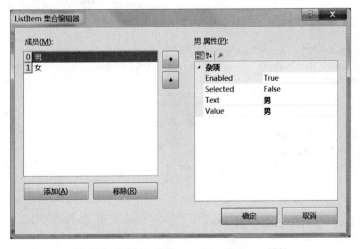

图 1-9 设置 RadioButtonList 的 itsms 属性

❺ 切换到页面的"源"视图,看到 HTML 标记。在<head>标记中加入以下 CSS:

```
<head id="Head1" runat="server">
<meta http-equiv="Content-Type" content="text/html; charset=utf-8"/>
    <title></title>
    <style type="text/css">
        .tableStyle
        {
            width:300pt;
            border:1pt;
        }
        .headerStyle
        {
            text-align:center;
            font-size:20pt;
        }
        .cellStyle
        {
            width:150pt;
        }
        .auto-style1
        {
            width: 150pt;
            height: 36px;
        }
    </style>
</head>
```

说明:

① HTML 在第 3 章将有简单的介绍。

② CSS 用于控制网页显示格式,如字体大小、颜色、位置等。网页中使用 CSS 有引用外部 CSS 文件、<head>中加入<style>标记、在 HTML 的开始标记中使用 style 属性设置,CSS 的用法见第 3 章。

❻ 在 HTML 标记中,通过 class 属性应用❺中设置的 CSS,<form></form>间的 HTML 修改如下。

```
<form id="form1" runat="server">
  <div>
    <table border="1" class="tableStyle">
    <tr>
     <td colspan="2" class="headerStyle">
       <asp:Label ID="Label1" runat="server" style="text-align: center" Text="学生基本情况"></asp:Label>
     </td>
    </tr>
    <tr>
    <td class="cellStyle">学号</td>
    <td class="cellStyle">
       <asp:TextBox ID="txtStudent_no" runat="server" Width="270px"></asp:TextBox>
    </td>
    </tr>
    <tr>
     <td class="cellStyle">姓名</td>
     <td class="cellStyle">
```

```
            <asp:TextBox ID = "txtStudent_name" runat = "server" Width = "270px"></asp:TextBox>
          </td>
        </tr>
        <tr>
          <td class = "auto-style1">性别</td>
          <td class = "auto-style1">
            <asp:RadioButtonList ID = "RdoGender" runat = "server" RepeatDirection = "Horizontal">
              <asp:ListItem>男</asp:ListItem>
              <asp:ListItem>女</asp:ListItem>
            </asp:RadioButtonList>
          </td>
        </tr>
        <tr>
          <td class = "cellStyle">
            <asp:Button ID = "Button1" runat = "server" Text = "提交" OnClick = "Button1_Click" />
          </td>
          <td class = "cellStyle">
            <asp:Button ID = "Button2" runat = "server" Text = "重置" OnClick = "Button2_Click" />
          </td>
        </tr>
      </table>
      <asp:Label ID = "Label2" runat = "server" Text = "Label"></asp:Label>
    </div>
</form>
```

❼ 在页面的"设计"视图中，双击"提交"命令按钮，在 Button1_Click 事件中增加代码：

```
string result = "学号：" + txtStudent_no.Text + "<br>" + "姓名：" + txtStudent_name.Text
    + "<br>" + "性别：" + RdoGender.Text;
Label2.Text = result;
```

❽ 运行 1_1.aspx。在 Google Chrome 浏览器的地址栏中输入 http://localhost:81/example/chapter1/1_1.aspx，填写完内容后，单击"提交"按钮，显示结果如图 1-10 所示。

说明：该程序也可以在 VS 中通过单击工具栏上的三角按钮直接运行。书中示例是在 Google Chrome 浏览器中运行的。由于 1_1.aspx 放在 chapter1 文件夹中，故运行程序时需要在路径中加上文件夹名。

图 1-10　程序运行结果

1.6.2　MVC 模式

MVC 开发模式包括 3 个模块，即 Model（模型）、View（视图）、Controller（控制器），由于每个模块不止一个，故开发过程中通常以 Models、Views 和 Controllers 表示。Models 负责与数据库交互，Views 负责页面的展示，Controllers 负责处理页面的请求。

1.6.3　使用原生态 Web 表单模式

原生态开发模式就是服务器端构造表单的代码与最终呈现给客户端的代码完全一致，用户更加方便地使用 CSS、JS 代码以及 jQuery、Ajax 来控制页面的显示效果和数据异步加载。

例 1-2 在原生态表单下，建立图 1-10 所示的表单，实现与例 1-1 相同的功能。

❶ 在"解决方案资源管理器"中，右击文件夹 chapter1，选择快捷菜单中的"添加"→"HTML 页"，在"指定项名称"中输入 1_2.html，单击"确定"按钮。

说明：在 1_2.html 的"源"视图中，可以看到 VS 已经自动编写了 HTML：

```html
<!DOCTYPE html>
<html xmlns="http://www.w3.org/1999/xhtml">
<head>
<meta http-equiv="Content-Type" content="text/html; charset=utf-8"/>
    <title></title>
</head>
<body>
</body>
</html>
```

❷ 将 1_1.aspx 的<style></style>粘贴到 1_2.html 的<head>中，在<body></body>间增加以下 HTML 代码，建立一个表单：

```html
<div>
<form id="ff" method="get" action="1_2.aspx">
<table>
  <tr><td colspan="2" class="tdTitle">学生基本情况</td></tr>
  <tr><td>学号</td><td><input type="text" name="sno"/></td></tr>
  <tr><td>姓名</td><td><input type="text" name="sname"/></td></tr>
  <tr><td>性别</td><td><input type="radio" name="gender" value="男"/>男<input type="radio" name="gender" value="女"/>女</td></tr>
  <tr><td><button type="submit" name="btnSave" id="btnSave">提交</button></td><td><button type="reset">重置</button></td></tr>
</table>
</form>
</div>
```

说明：<form>标记常用的属性有 method 和 action。method 是表单数据提交的方式，有 get 和 post 两种方式。action 指定处理表单数据的程序，上面指定的是 1_2.aspx，如果只是完成数据逻辑运算，而不显示数据，处理程序也可以指定为"一般处理程序"（扩展名是*.ashx）。

❸ 在"解决方案资源管理器"中，右击文件夹 chapter1，选择快捷菜单中的"添加"→"Web 窗体"命令，在弹出的对话框中的"指定项名称"中输入 1_2.aspx，单击"确定"按钮，系统会自动打开 1_2.aspx。切换到"源"视图。在"解决方案资源管理器"中单击 1_2.aspx 文件名前面的 ▷ 按钮，展开后找到 1_2.aspx.cs，单击打开 1_2.aspx.cs，在 Page_Load 中写入以下代码：

```csharp
protected void Page_Load(object sender, EventArgs e)
{
    string sno = Request["sno"];
    string sname = Request["sname"];
    string gender = Request["gender"];
    string result = "学号：" + sno + "<br>" + "姓名：" + sname + "<br>" + "性别：" + gender;
    Response.Write(result);
}
```

❹在浏览器的地址栏中输入：http://localhost:81/example/chapter1/1_2.html。

说明：

① Request 和 Response 是 ASP.NET 提供的内部对象。Request 用于从客户端收集数据，常见用法是 Request["x"]，其中 x 与表单中控件的 name 相对应，如 Request["sno"] 得到表单中 name="sno" 文本框中的数据。

从一个页面跳转到另一个页面时，有时需要以超链接的方式在页面间传递数据，如 <a href="a.aspx?a=1">click me</a>，单击 click me 后，页面转向 a.aspx，变量 a 的值也传递到该页面。在 a.aspx 中，可以使用 Request["a"] 接收上一个页面传递过来的值。

Response 对象用于服务器向客户端输出信息，该对象最常用的方法是 Write() 和 Redirect()。通过 Response.Write() 输出信息时，可以使用 HTML，如 Response.Write("<font color=red>北京大学</font>")，此代码的结果是输出红色字的"北京大学"。

② 与例 1-1 不同之处在于，单击"提交"按钮后，浏览器会切换到另一个新的页面，不能将 Response.Write(result) 输出的 result 与表单出现在同一个页面。例 1-1 的表单提交后，表单中的数据能够维持 ViewState（视图状态），是通过在带有 <form runat="server"> 控件的每个页面上放置一个隐藏域实现的。维持 ViewState 是 ASP.NET Web Forms 的默认设置。如果不想维持 ViewState，可在 .aspx 页面顶部包含指令 <%@ Page EnableViewState="false" %>，或者向任意控件添加属性 EnableViewState="false"。

③ 例 1-2 这种提交表单的方式与 ASP.NET 之前出现的 ASP 基本相似。原生态表单下要维持 ViewState，可以使用 jQuery（jQuery 是一个 JavaScript 库），例 1-3 简单地说明了这一过程。

要使用 jQuery，需要将 jQuery 的 jquery.min.js 引用到网页中。引用时涉及网页与 jQuery 库文件的相对位置关系，图 1-11 是本书中文件间的关系。

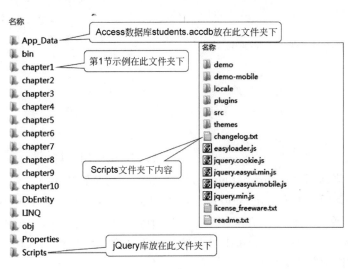

图 1-11　本书中文件间相对关系

例 1-3 以例 1-1 为例，使用 jQuery 建立原生态的表单。

❶ 在"解决方案资源管理器"中，右击文件夹 chapter1，选择快捷菜单中的"添加"→"HTML 页"命令，在弹出的对话框中的"指定项名称"中输入 1_3.html，单击"确定"按钮。

❷ 在 1_3.html 中输入以下代码：

```html
<body>
<script type="text/javascript" src="../Scripts/jquery.min.js"></script>
<div>
<form id="ff" method="get">
<table class="tableStyle">
<tr><td colspan="2" class="headerStyle">学生基本情况</td></tr>
<tr><td class="cellStyle">学号</td><td class="cellStyle"><input type="text" name="sno" /></td></tr>
<tr><td class="cellStyle">姓名</td><td class="cellStyle"><input type="text" name="sname" /></td></tr>
<tr><td class="cellStyle">性别</td><td class="cellStyle"><input type="radio" name="gender" value="男" checked />男
<input type="radio" name="gender" value="女" />女</td></tr>
<tr><td class="cellStyle"><button id="btnSave">提交</button></td>
<td class="cellStyle"><button type="reset">重置</button></td></tr>
</table>
</form>
<div id="content"></div>
</div>
<script>
    $(document).ready(function () {
        $("#btnSave").click(function () {
            var sendData = $("#ff").serialize();
            $.get("1_3.aspx", sendData, function (returnData, status) {
                $("#content").html(returnData);
            });
            return false;     //此句不可少
        })
    })
</script>
</body>
```

说明：

① <script src="">引用 jQuery 库 jQuery.min.js，该引用也可以放在<head></head>间，引用时注意当前网页文件与 jQuery.min.js 文件间的位置关系。

② <form>标记中不指定 action 属性，处理表单的程序文件由 jQuery 指定。

③ 第 11 行命令按钮是放在<form></form>间，故 return false; 不能省略；否则单击"提交"按钮后表单会刷新，表单上填写的数据将清空。如果将<button>放在<form></form>外，则不需要该语句。

④ <script></script>间是使用 JavaScript 或 jQuery 编写的客户端脚本。

$()会返回一个新的 jQuery 对象实例，$()的写法在 JavaScript 库中很常见，一个页面中如果使用了多个这样的库，可能会发生冲突，$()可以替换成 jQuery()。

使用$(document).ready()方法,jQuery使预定的DOM(Document Object Model)加载完毕后,自动调用指定的函数,不必等待页面中图像的加载。

⑤ 代码 var sendData = $("#ff").serialize();将表单数据序列化,生成URL编码文本字符串,若在该语句后加入alert(sendData);,查看到的结果见图1-12。左边是输入的内容,右边是alert()弹出的对话框。

图1-12 1_3.html程序输入内容及表单序列化结果

由图1-12可看出,serialize()方法序列化后URL编码格式为xx=**&yy=**,xx和yy是表单中各控件的name属性值,**是各控件输入的值。URL编码后中文显示的虽然是乱码,但不影响1_3.aspx中通过Request对数据的接收。

⑥ $.get()方法使用HTTP GET请求从服务器加载数据,其参数说明如图1-13所示。

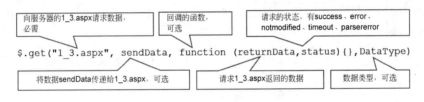

图1-13 参数说明

本示例将输入的内容发送到服务器,服务器响应后将接收到的信息输出。服务器收到客户端传递过来的数据,如查询信息,需要从数据库中执行查询,然后将查询结果返回给客户端。

⑦ $("#content").html(returnData),使用jQuery的html()方法,将返回的数据输出至<div id="content"></div>间。

⑧ <script></script>中的return false;用于阻止表单提交数据,防止页面转到一个新的网页。如果将<button id="btnSave">提交</button>放在<form></form>的外面,可以取消此语句。

❸ 在chapter1文件夹下,新建立一个名为1_3.aspx的Web窗体文件,在1_3.aspx.cs文件的Load事件中输入以下代码:

```
string sname = Request["sname"].ToString();
string sno = Request["sno"].ToString();
string gender = Request["gender"].ToString();
Response.Write("学号:" + sno + "<br/>");
Response.Write("姓名:" + sname + "<br/>");
Response.Write("性别:" + gender + "<br/>");
```

例 1-3 通过使用 jQuery,在不刷新表单的情况下提交表单。

jQuery 在本书的第 12 章将有详细的介绍。

1.6.4 三层架构

使用上述 3 种开发模式编写程序代码时,为了使代码可重用以及便于维护,书写代码时一般采用三层架构,这里的"层"是对代码的一种逻辑划分,并不一定要三层,如果系统很简单,只有一个页面,那一层就可以;如果系统很复杂,也可能是 n 层。三层架构指的是数据访问层(DAL)、业务逻辑层(BLL)、用户界面层(UIL)。数据访问层负责与数据库连接、建立数据访问对象等,业务逻辑层将数据访问层中的数据加工和处理,用户界面层将业务逻辑层中加工处理后的数据以各种形式展示出来。

本书第 6 章详细地介绍了使用三层架构访问 SQL Server 的方法。

习题和思考

(1) 简述 Web 控件模式和原生态表单模式各自的优点与缺点。

(2) 原生态模式下,通过 jQuery 建立一个简单的表单,程序运行后在文本框中输入要问候的话,单击"问候"按钮,在当前页面显示出输入的问候语,如图 1-14 所示。(提示:可以模仿例 1-3,也可以在 HTML 文件中通过 $("#文本框控件的 id").val()得到文本框中输入的值,然后直接输出。)

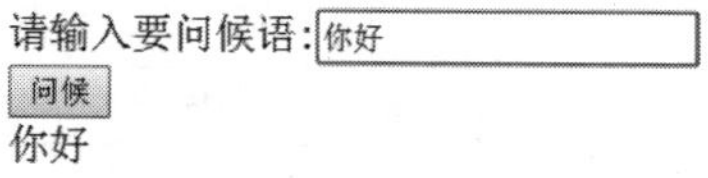

(3) 练习申请云,并在云上搭建 ASP.NET 运行平台。

图 1-14 程序运行结果

(4) 新建立一个 Web 窗体,窗体上拖曳一个命令按钮 Button1,在"设计"视图下,双击该按钮,写入以下代码(第 1 行不用输入,由 Visual Studio 自动生成):

```
protected void Button1_Click(object sender, EventArgs e)
{
  for (int i = 0; i < 5; i++)
  {
    Response.Write("北京科技大学<br/>");
  }
}
```

运行该 Web 窗体,单击 Button1 命令按钮后,查看该页面的源代码,比较源代码与自己输入代码的差异。

提示:运行网页后,查看源代码的方法如下。

- 在 IE 浏览器中右击空白处,选择快捷菜单中的"查看源"命令。
- 在 Google Chrome 浏览器中右击空白处,选择快捷菜单中的"查看网页源代码"命令。

数据库基础知识

(1) 数据库管理系统。

(2) SQL Server 简介。

(3) SQL Server 中常用的 SQL。

2.1 数据库管理系统

数据库管理系统(DataBase Management System,DBMS)是一个软件系统,主要用来定义和管理数据库,处理应用程序和数据库间的关系。数据库管理系统是数据库系统的核心部分,它建立在操作系统之上,对数据库进行统一管理和控制。简单地说,DBMS 就是帮助我们建立、管理和维护数据库的软件系统。其主要功能如下。

1. 描述数据库

DBMS 能够提供数据描述语言(Data Description Language,DDL),描述数据库的逻辑结构、存储结构和保密要求等,通过它能够方便地建立数据库和定义数据库结构。

2. 操作数据库

DBMS 能够提供数据操作语言(Data Manipulate Language,DML),通过它能够方便地对数据库进行查询、插入、修改和删除等操作。

3. 管理数据库

DBMS 能够提供对数据库的运行和管理功能,保证数据的安全性、完整性和一致性,能够控制并发用户对数据库的访问,管理大量数据的存储。

4. 维护数据库

数据库管理系统能够提供数据的维护功能,如数据的导入、导出、数据转换、备份、故障恢复和性能监视等。

平常所说的会使用某种数据库，实际上指的是会使用该数据库管理系统。常见的DBMS从规模上划分，可分为桌面型数据库管理系统和网络型数据库管理系统。桌面型数据库管理系统有 Access、Visual Foxpro 等；网络型数据库管理系统有 Oracle、SQL Server、Informix、Sybase、DB2 等。

2.2 数据库

数据库(DataBase,DB)是存储在计算机内有组织、可共享的数据集合。数据库中的数据按照一定的数据模型组织、描述和存储，其特点是具有较小的冗余度、较高的独立性和可扩展性，并且数据库中的数据可供各种合法的用户使用。

针对不同行业、不同研究内容，可以使用各种 DBMS 建立不同的数据库，如使用 SQL Server 建立人事管理数据库、使用 Access 建立学生成绩管理数据库。

2.3 关系数据库

目前大多数数据库管理系统是基于关系模型的关系数据库，由于关系数据库建立在严格的数学基础上，并且结构简单、使用方便，因而得到广泛的应用。关系数据库中数据的基本结构是表，即数据是按行、列有规则地排列、组织。一个数据库中可以有一个或多个表。

关系数据库中涉及的主要概念有以下几个。

(1) 关系：一个关系在逻辑上对应一个按行、列排列的二维表。

(2) 字段：又称为属性，表中的一列称为表的一个属性，其反映的是研究对象某一方面的特性。

(3) 记录：又称为元组，是表中的每一行。

(4) 主键：在表中能唯一地标识记录的一个字段或字段的集合。

(5) 外键：表 A 中的某一字段，在该表中虽然不是主键，或是作为主键的一部分，但该字段在表 B 中是主键，那么这个字段在表 A 中称为外键。

在表 2-1 中，"学号""姓名""性别""籍贯""电话"为字段，有 3 条记录。表中字段"学号"为主键，因为如果知道了"学号"，就能唯一地确定一条记录。"姓名"不能作为主键，因为有重名，不能根据姓名唯一地确定一条记录。

表 2-1 学生表

学号	姓名	性别	籍贯	电话
9801001	张三	男	北京	8233444
9801002	李四	女	上海	6578432
9901001	张三	女	天津	9876543

表 2-2 中"学号"和"课号"都不是主键，因为一个学生不可能只学一门课，而每门课也不可能只有一个学生学习，主键应该是"学号"+"课号"，用字段的集合作为主键。表 2-2 中"学号"不是主键(只是主键的一部分)，但"学号"在表 2-1 中是主键，故表 2-2 中"学号"是外键。表 2-1 和表 2-2 两个表通过"学号"关联，可以组合生成表 2-3。

表 2-2 成绩表

学号	课号	成绩
9801001	K001	80
9801002	K001	85
9801001	K002	79
9801002	K002	90

表 2-3 学生表和成绩表左关联的结果

学号	姓名	性别	籍贯	电话	课号	成绩
9801001	张三	男	北京	8233444	K001	80
9801001	张三	男	北京	8233444	K002	79
9801002	李四	女	上海	6578432	K001	85
9801002	李四	女	上海	6578432	K002	90
9901001	张三	女	天津	9876543	NULL	NULL

数据库中表与表之间的关联方式有3种,即左关联、右关联和内部关联。将表 2-1 当作左表,表 2-2 当作右表。左关联表示关联后生成的表中,包含表 2-1 中的所有记录及表 2-2 中"学号"与表 2-1 中"学号"相等的记录;内部关联表示关联生成的表中,包含表 2-2 中"学号"与表 2-1 中"学号"相等的记录;右关联表示关联生成的表中,包含表 2-2 中的所有记录及表 2-2 中"学号"与表 2-1 中"学号"相等的记录。由此可以看出,表 2-3 是表 2-1 与表 2-2 通过"学号"左关联的结果,表中 NULL 表示空值。

2.4 数据库系统

数据库系统(DataBase System,DBS)是由文件系统演变而来的,数据库系统中的数据不是针对某个具体的应用程序而设计的,而是面向全局的应用。一个数据库中可以包括表、视图、存储过程、各字段的属性、规则等。数据库系统主要特点如下。

1. 统一管理的结构化数据

数据库系统中的数据是有结构的,由 DBMS 统一管理。在设计数据库结构时,在调查研究的基础上,要充分考虑整个系统的数据结构,不是以某个具体的应用程序为依据,既要描述数据,也要描述数据之间的关系。

2. 数据冗余度小

合理的数据库系统要尽量减小数据的冗余度。在一个系统中,可能涉及多个用户,不同的用户根据不同的需要访问不同的数据子集。减小冗余度的好处主要有两个:一个是节约数据存储的空间;另一个是避免数据的不一致性。

3. 数据共享

数据库系统中的数据能够为系统中所有合法的用户共享使用,也可为系统中各类应用程序共享使用。

4. 数据的独立性

数据的独立性是指在数据库中的数据及数据的组织与应用程序无关,也就是说,数据库中的数据发生改变时,应用程序不用发生改变。数据的独立性包括逻辑独立性和物理独立性两个方面。

2.5 数据库中表间的关系

1. 表和表间的关系

表和表间的关系有 1：m(1 对多)、m：n(多对多)和 1：1(1 对 1)。在一个数据库中,一般都包含有多个相关的表,数据是存放在表中的,数据库可以看作表的容器。表 2-1 和表 2-2 通过"学号"相关联。表 2-1 中的一条记录,如"学号＝9801001"的记录,与表 2-2 中的两条记录有关联,二者的关系为一对多的关系,表示为 1：m。

关系数据库不能处理 m：n 关系,一般要将一个 m：n 折分成两个 1：m 来处理。对于 1：m 关系的两个表,在建立数据库时,一般是不能将其合并为一个表,如表 2-1 和表 2-2 的关系是 1：m,就不能将其合并为一个表。

2. 如何建立合理的数据表

假设要开发一个学生管理的信息系统,其中涉及学生基本信息和学生成绩信息。建立数据表前,经过系统调查,总结出系统中涉及的信息有学号、姓名、性别、籍贯、电话、课号、成绩。根据实际情况,考虑以下两种建立数据库表的模式。

关系模式一：

学生(学号、姓名、性别、籍贯、电话、课号、成绩)

关系模式二：

学生(学号、姓名、性别、籍贯、电话)
成绩(学号、课号、成绩)

如果采用关系模式一建立数据库中的表时,输入数据后,表的内容见表 2-3。如果某一同学要学习 50 门课,该同学的学号、姓名、性别、籍贯、电话等信息就要重复 50 次,这样会造成数据冗余。数据冗余最可怕的后果是造成数据的不一致性,在表 2-3 中输入成绩时,电话号码输入错误,这样在查询张三的电话号码时会发现有多个。

如果采用关系模式二,在一个数据库中建立两个表,一个表是"学生",另一个表是"成绩",二者可通过"学号"相联系。"学生"中"学号"是主键,因为知道"学号",就能确定唯一记录。"成绩"中"学号"和"课号"合起来是主键,二者中的任何一个都不能确定唯一记录,但合起来可以确定唯一记录。将一个大的表,分解成表 2-1 和表 2-2 两个小表后,明显地减小了数据冗余。

在数据库管理系统开发前,一定要先建立好数据库。合理的数据库是建立在对项目充分调查研究和分析基础上的。在完成一个项目时,切忌没有进行系统分析就编写程序。如果程序快开发完毕,再修改数据库结构,已编好的程序大都要进行修改,这将大大影响开发效率。

下面采用关系数据库的规范化理论,针对上面的例子进一步加以阐述。这些理论是设计数据库时用来进行数据分析的一种方法,有助于建好数据库。标准化的过程是一系列渐近的规则集,一般称为范式。由于范式是渐近的规则集,所以应该先用第一范式,再用第二范式,依此类推。在实际程序开发中,设计数据库时一般满足第三范式即可。

1) 第一范式(1NF,即 First Normal Form)

通俗地讲,就是表中的各字段是最小单位,不可分割。

很显然表 2-1~表 2-3 都满足 1NF。如果是表 2-4,就无法满足 1NF。由于"电话"不是最小单位,可进一步拆分,建立数据库时,要将"电话"字段改为"家庭电话"和"办公室电话"才能满足 1NF。

表 2-4 不满足 1NF 的表

学号	姓名	性别	籍贯	电 话		课程名	成绩
				家庭	办公室		
9801001	张三	男	北京	8233444	8233225	英语	80
9801001	张三	男	北京	8233444	8233225	政治	78

2) 第二范式(2NF)

某关系在满足 1NF 的前提下,非主键(除主键外的字段)是由主键的全部而不是部分决定的,则此关系满足 2NF。

关系模式一满足 1NF,但不能满足 2NF。原因是关系模式一中主键是"学号"+"课号",其余的都是非主键。在非主键中,"姓名""性别"等可由主键的部分"学号"即可确定。

通过分析可知,关系模式二能够满足 2NF。

3) 第三范式(3NF)

某关系如果能够满足 2NF,并且在非主关键字段中,各字段间互不依赖,即除了主关键字外的所有字段,各字段间互不相关,则此关系满足 3NF。

通过分析可知,关系模式二能够满足 3NF。

本书主要使用 SQL Server 数据库管理系统,将逻辑设计转变为物理设计。

2.6 SQL Server 的基本操作

微软在 https://www.microsoft.com/zh-cn/download/details.aspx?id=29062 提供了 Windows 32 位和 64 位免费的 SQL Server 2012 Express。安装成功后,通过 SQL Server Management Studio 可以可视化地建立数据库、编辑数据表、完成数据的导入和导出等操作。在安装过程中,本书选择的 SQL Server 的身份验证方式是"混合模式(SQL Server 身份验证和 Windows 身份验证)",其界面如图 2-1 所示,用户名为 sa,密码为 3500。

2.6.1 建立数据库 students

❶ 运行 SQL Server Management Studio 软件,登录成功后右击"数据库",选择快捷菜单中的"新建数据库"命令,在弹出的"新建数据库"对话框,输入数据库名 students,单击"确定"按钮。

❷ 单击"数据库"下新建立的数据库 students 前面的"+"按钮,展开 students 数据库,

选中该库下的"表"并右击,选择快捷菜单中的"新建表"命令,建立图 2-2 所示的数据表,id 字段的属性设置为"标识字段"。右击该字段,选择快捷菜单中的"设置主键"命令。

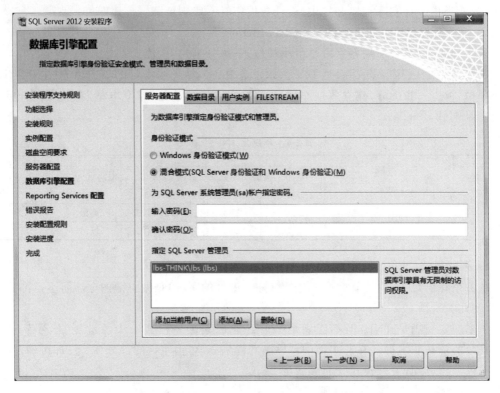

图 2-1　安装 SQL Server 时身份认证的界面

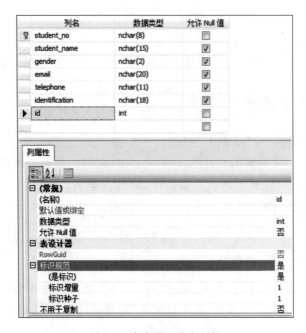

图 2-2　表中设置字段属性

❸ 关闭新建立的表,弹出是否要保存表的提示框,单击"是"按钮,输入表名 student 并单击"确定"按钮。

❹ 重复以上操作,建立 score、course、teacher 表,表中字段见图 2-3～图 2-5。

列名	数据类型	允许 Null 值
student_no	nchar(8)	□
course_no	nchar(8)	□
score	decimal(5, 1)	☑
id	int	□

图 2-3　score 表结构

列名	数据类型	允许 Null 值
course_no	nchar(8)	□
course_name	nchar(20)	☑
course_address	nchar(30)	☑
course_time	nchar(16)	☑
teacher_ID	nchar(10)	☑
id	int	□

图 2-4　course 表结构

列名	数据类型	允许 Null 值
teacher_ID	nchar(10)	□
teacher_name	nchar(10)	☑
teacher_gender	nchar(2)	☑
teacher_phone	nchar(11)	☑
id	int	□

图 2-5　teacher 表结构

上述 4 个表中,字段 student_no、student_no＋course_no、course_no、teacher_ID 均是主键,每个表中都有字段 id,其数据类型是整型,设置标识规范属性为"是",标识的种子是 1,标识增量是 1,表示每增加一条记录,该字段的值从 1 开始自动加 1。

默认情况下,SQL Server 不允许修改已经建好的表结构。如果要修改,需要在 SQL Server 中选择"工具"→"选项"命令,打开"设计器",去除"阻止保存要求重新创建表的修改"复选框。

2.6.2　建立数据库关系图

目前已经为数据库 students 建立了 4 个表,这 4 个表间通过主键和外键相互关联。如果将 student 表中 student_no 为 1 的记录删除或者将 student_no 由 1 改为 2,其相关联的表 score 也应该做出调整。这些变化也可以通过前台数据库实现,但如果建立数据库关系图,可由数据库管理系统自动完成。

❶ 右击"数据库"→"数据库关系图",在弹出的快捷菜单中选择"新建数据库关系图"命令。

❷ 将 student、teacher、course、score 4 个表增加到"新建数据库关系图"中,如图 2-6 所示。

❸ 在图 2-6 中,选中 student 表 student_no 字段前的 ▮ 图标拖曳至 score 表 student_no 字段前的 ▮ 图标处松开鼠标键,出现图 2-7 所示的确认这两种表关系的对话框。

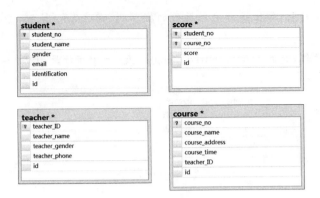

图 2-6　建立数据库关系

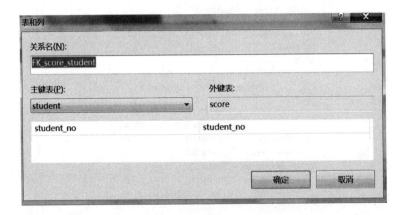

图 2-7　确认表间的关系

❹ 用同样的方法,建立表 teacher 与表 course 的关系,表 course 与表 score 的关系。建立后,图 2-6 变成图 2-8。

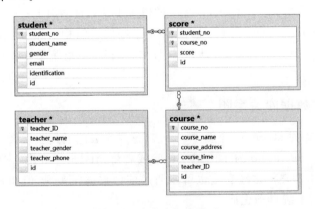

图 2-8　库中表间的关系

图 2-8 中各表之间的关联线表示这几个表间的关系是 $1:m$,选中关联线可以设置其属性。下面是属性"INSERT 和 UPDATE 规范"的说明(以 student 和 score 间的关联线为例)。

选中该关联线,在属性"INSERT 和 UPDATE 规范"前单击,打开其内容,如图 2-9 所示。更新和删除规则下有不执行任何操作、级联、设置 Null、设置默认值。设置"更新规则"为"级联"时,如果修改 student 表中 student_no 的值,score 表中相对应的 student_no 自动修改;设置"删除规则"为"级联"时,student 表中删除某条记录时,score 表中那些与 student 表中 student_no 相同的记录会自动删除。

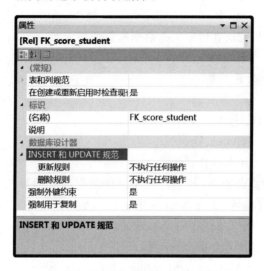

图 2-9　设置更新规则和删除规则

强制外键的约束:表示子表中插入数据或者更新数据时,外键值必须参照与其关联的父表中主键的值。例如,student 和 score 两个表,student 是主表,score 是子表,在 score 表中输入 student_no 记录时,输入的值必须在 student 表中存在,否则输入失败;更新 score 中如 student_no 的值,也必须参照 student 中 student_no 的值。

通过设置图 2-9 中"INSERT 和 UPDATE 规范"及"强制外键约束"选项,SQL Server 保证表中数据的有效性。

2.6.3　在数据库中增加记录

运行 SQL Server Management Studio,登录成功后,展开"数据库"前面的"＋",进一步展开 students 库前面的"＋"再进一步展开"表"前面的"＋",右击 student 表,在弹出的快捷菜单中选择"编辑前 200 行"命令,输入记录值。上述各表中的 id 字段,增加记录时会自动增加,用户不能通过程序代码或手工操作为该字段赋值。

2.6.4　权限设置

通过网页操作数据库,在 SQL Server 2012 Express 中必须设定数据库指定操作权限;否则程序在打开数据库时会出现错误。设置权限的步骤如下。

❶ 右击数据库 students,在弹出的快捷菜单中选择"属性"命令,在"权限"中单击"搜索"按钮,出现"选择用户或角色"对话框。

❷ 单击"浏览"按钮,选中 public,单击"确定"按钮,返回"数据库属性"对话框。

❸ 通过勾选"public 的权限"的列表框中的"插入""删除""更改""选择"等复选框以授予相应的权限。

2.7 关系数据库标准语言——SQL

SQL(Structured Query Language,结构化查询语言)是一种在关系数据库中定义和操作数据的标准语言,1974 年由 Boyce 和 Chamberlin 提出,当时称为 SEQUEL 语言,1976 年由 IBM 公司的 San Jose 研究所在研制关系数据库管理系统 System R 时,修改为 SEQUEL 2,也就是目前十分流行的 SQL 语言。

SQL 语言通常分为以下 4 类。

(1) 查询语言(Select)。

(2) 操作语言(Insert、Update、Delete)。

(3) 定义语言(Create、Alter、Drop)。

(4) 控制语言(Commit、RollBack)。

SQL 语言的最大特点是直观、易学。在不同的数据库管理系统中,SQL 基本上是相同的,但在个别地方会稍有差异,如在 Access 中通配符是"*",而在 SQL Server 中通配符是"%",这一点在编写数据库应用程序时要特别注意。

在 SQL Server 中运行 SQL 的方法有以下几个。

(1) 运行 SQL Server Management Studio,登录 SQL Server,选中 students 数据库,选中数据库 students。

(2) 单击工具条中的"新建查询"按钮,出现输入 SQL 的对话框。本节的 SQL 都可以在此输入练习,输入后的 SQL 也可以文本的形式保存下来,文件的扩展名为.sql。

(3) 单击工具条上的"执行"按钮,运行 SQL。如果 SQL 书写正确,应出现运行结果;如果不正确,系统会提示出现的错误。

2.7.1 单表数据查询

语句结构:

select...from...where...

(1) 查询 student 表中全部记录,要求列出全部字段:

select student.* from student

语句中"*"表示所有字段。如果查询的内容来自一个表,可以省略"表名.",写作*,上边的语句可以写作 select * from student。

(2) 从 student 表中查询学生全部记录,包括 student_no、student_name、gender 这 3 列,并将显示结果的标题改为"学号""姓名""性别",如图 2-10 所示。语句为:

select student.student_no as 学号,student.student_name as 姓名,
student.gender as 性别 from student

由于显示的字段来源于一个表 student,故上述 select 语句也可写为:

	学号	姓名	性别
1	41321059	李孝诚	男
2	41340136	马小玉	女
3	41355062	王长林	男
4	41361045	李将寿	男
5	41401007	鲁宇星	男
6	41405002	王小月	女
7	41405003	张晨露	女

图 2-10 查询结果

```
select student_no as 学号, student_name as 姓名, gender as 性别 from student
```

select 语句中用 as 后面的内容,表示查询结果的标题。

(3) 从 student 表中查询 gender 是"女"的记录,显示 student_no、student_name、gender 这 3 列,语句为:

```
select student_no as 学号, student_name as 姓名, gender as 性别 from student where gender = '女'
```

本语句中"where"用于条件查询,多条件时用 and(与条件)或者 or(或条件)连接。

(4) 从 student 表中查询姓"王"的学生,列出全部字段,语句为:

```
select * from student where student_name like "王%"
```

语句中"%"表示通配符,表示多个字符,有的 DBMS 如 access,要写为"*"。SQL Server 中下边语句完成与上边语句相同的功能:

```
select * from student where substring(student_name,1,1) = '王'
```

substring 是字符串截取函数,substring(student_name,1,1)中第 1 个参数 1,表示截取 student_no 的起始位置,第 2 个参数 1 表示截取 student_no 的长度。

(5) 如果 student 表中没有 gender 列,根据身份证号判断学生的性别、年龄,显示"姓名""性别""年龄"(计算年龄只考虑年,不考虑月和日)3 列:

```
select student_name as 姓名, iif(substring (identification,17,1)%2 = 0,'女','男') as 性别, year(getdate()) - substring (identification,7,4) as 年龄 from student
```

getdate()取出系统当前日期和时间,iif()用于条件判断,%是求余运算符。低于 SQL Server 2012 的版本,没有 iif 函数,该语句可改为:

```
select student_name as 姓名, case when substring(identification,17,1)%2 = 0 then '女' else '男' end, year(getDate())-substring(identification,7,4) as 年龄 from student
```

(6) 查询学号是"41321059"或者"41340136"的全部记录:

```
select * from student where student_no in('41321059', '41340136')
```

语句中使用 in 运算符。多个值之间要用","分开。也可以写为:

```
select * from student where student_no = '41321059' or student_no = '41340136'
```

显然,如果查询的是多个值,使用 in 更简练。

(7) 查询学号在 41321059~41340136 间的记录:

```
select * from student where student_no between '41321059' and '41340136'
```

查询在某个范围,书写条件时,可以使用 between…and…。

(8) 列出 score 表中的课号:

```
select course_no as 课号 from score
```

该语句执行后,由于一门课可以有多个学生学习,故出现许多重复的课号,要避免此种情况,应使用语句:

```
select distinct course_no as 课号 from score
```

distinct 用于过滤重复的记录。

2.7.2 排序

语句结构：

select...from ...where...order by...

(1) 查询 student 表中全部记录，且按姓名排序：

```
select * from student order by student_name
```

从执行结果可以看出，默认情况下，排序方式是升序(按照汉字的拼音字母由 a 到 z)，该语句相当于"select * from student order by 姓名 asc"。asc 的英文单词是 ascend，是上升之意，但语句中只能写 asc，不能写 ascend。如果要按姓名降序排列，语句为：

```
select * from student order by student_name desc
```

desc 表示 descend。

(2) 查询 student 表中全部记录，先按 gender 升序排序，然后再按 student_name 降序排序：

```
select * from student order by gender asc, student_name desc
```

(3) 从 student 表中随机挑选 1 条记录：

```
select top 1 * from student order by newID()
```

newID()返回一个 uniqueidentifier(全局唯一的标识)值，top 1 选取排序后的第 1 条记录。

(4) 从 student 表中查询女生记录，将查询结果输出到一个表文件 result 中：

```
select * into result from student where gender = '女'
```

2.7.3 统计和分组

语句结构：

select...from ...where...group by...having...

在某些情况下，常常将一些筛选出的数据作一些分类，而将数据分成若干集合，如将所有学生的成绩按学号加以分类，再对每一个集合进行统计分析。不使用 group by 所筛选出的数据也是以集合的形式存在，只是它们自成一个集合而非数个集合。需要说明的是，所筛选出的数据集合可能包含多笔数据、一笔数据或无数据。

常用的集总函数(专门为分析 GROUP BY 之后的每一个集合数据而设计的一些函数)有以下几个。

1. count()函数

语法结构：

```
count([all|distinct expression]|[ * ])
```

功能：返回一个集合内所拥有的记录数。

参数说明：all 施用于所有的数值；distinct 返回唯一且非 Null 数值的个数；* 计算一个表格中所有记录的总笔数。

(1) 查询 student 表中有多少学生记录：

select count(*) as 学生总数 from student

(2) 按照性别，分组统计男女人数各多少，并且按照性别降序排序：

select gender as 性别,count(*) as 人数 from student group by gender order by gender desc

或者

select gender as 性别,count(*) as 人数 from student group by gender order by 性别 desc
select gender as 性别,count(*) as 人数 from student group by gender order by 1 desc

order by 排序的字段可以用表中原有的字段名，也可以使用 as 后的别名，还可以是要显示列的编号，如上面"性别"是要显示的第 1 列，排序时可以用 order by 1。

(3) 查询 score 表中有多少个不同的课号：

select count(distinct course_no) as 数量 from score

2. sum()函数

语法结构：

sum([all|distinct]expression)

功能：返回一个集合内所有数值或不同数值的总和，只能用于数字列，它会排除 NULL。

参数说明：all 用于所有的数值；distinct 表示 sum 返回不同数值的总和；expression 为一常数、列或函数。

(1) 查询成绩表中，学生的总成绩：

select sum(score) as 总成绩 from score

该语句将所有学生的成绩加在了一起。

(2) 根据 student_no，分组小计每个学生的总成绩，执行结果如图 2-11 所示：

select student_no as 学号,sum(score) as 总分 from score group by student_no

3. avg()函数

语法结构：

avg([all|distinct expression])

功能：返回一个集合内所有数值或不同数值的平均值，它会排除 Null。

(1) 根据 student_no，分组小计每个学生的平均成绩，执行后结果如图 2-12 所示：

select student_no as 学号,avg(score) as 平均分 from score group by student_no

	学号	总分
1	41321059	249.0
2	41340136	267.0
3	41355062	253.0
4	41361045	241.0
5	41401007	218.0
6	41405002	259.0
7	41405003	229.0

	学号	平均分
1	41321059	83.000000
2	41340136	66.750000
3	41355062	84.333333
4	41361045	80.333333
5	41401007	72.666666
6	41405002	86.333333
7	41405003	76.333333

图 2-11　分组小计总成绩　　　　图 2-12　分组小计平均成绩

如果图 2-12 中要求四舍五入,保留两位小数,可以输入以下代码:

select student_no as 学号, round(avg(score),2) as 平均分 from score group by student_no

如果采用 cast 函数,则运行结果如图 2-13 所示。代码如下:

select student_no as 学号, cast(avg(score) as decimal(10,2)) as 平均分 from score group by student_no

(2) 根据 student_no,分组小计每个学生的平均成绩,列出平均分大于 80 的学生,执行结果如图 2-14 所示。代码如下:

select student_no as 学号,avg(score) as 平均分 from score group by student_no having avg(score)>80

	学号	平均分
1	41321059	83.00
2	41340136	66.75
3	41355062	84.33
4	41361045	80.33
5	41401007	72.67
6	41405002	86.33
7	41405003	76.33

	student_no	平均分
1	41321059	83.000000
2	41355062	84.333333
3	41361045	80.333333
4	41405002	86.333333

图 2-13　运用 cast 函数　　　　图 2-14　列平均分大于 80 的学生

上边语句如果写为"select student_no as 学号,avg(score) as 平均分 from score group by student_no having 平均分大于 80",虽然在有的数据库管理系统中是正确的,但在 SQL Server 中是错误的。

图 2-14 相当于在图 2-12(或图 2-13)的基础上,选出平均分大于 80 的记录。相当于执行下边两条 select 语句的效果:

select student_no as 学号,avg(score) as 平均分 into result from score group by student_no
select * from result where 平均分>80

但如果写为"select student_no as 学号,avg(score) as 平均分 from score where avg(score)>80 group by student_no",则是错误的。对于分组后记录再进行筛选,要使用 having,而不能使用 where。select 语句中,只要有 having,其前边必须有 group by。在一条 select 语句中,select …from …where…group by…having…有可能同时出现,它们之间的关系可以用图 2-15 表示。

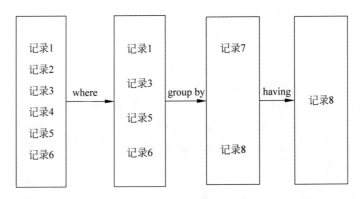

图 2-15　select 中 where、group by 和 having 关系示意图

4. max()函数

语法结构：

```
max(expression)
```

功能：返回一个集合内所有数值的最大值。

5. min()函数

语法结构：

```
min(expression)
```

功能：返回一个集合内所有数值的最小值。

	课号	最高分	最低分
1	k01	90.0	50.0
2	k02	90.0	62.0
3	k03	90.0	70.0
4	k04	60.0	60.0

图 2-16　查询最高分和最低分示例

例如，查询每一课程中的最高分和最低分，执行结果如图 2-16 所示：

```
select course_no as 课号,max(score) as 最高分,min(score) as 最低分 from score group by course_no
```

2.7.4　多表数据查询

如果查询的内容来自多个表，书写 select 语句时字段名前必须加上“表名.”，还要指定表间的关联关系(左关联、右关联、内部关联)。

(1) 查询显示课程名、上课时间、上课地点、任课老师。

“课程名”“上课时间”“上课地点”来自 course 表，“任课老师”来自 teacher 表，course 表与 teacher 表通过 teacher_id 关联，查询语句为：

```
select course.course_name as 课程名,course_address as 上课地点,course_time as 上课时间,
teacher.teacher_name as 任课教师 from course, teacher where course.teacher_id = teacher.
teacher_id
```

该 select 语句中使用 where 指明了 course 和 teacher 两个表间的关系是内部关联。多表查询用得最多的是内部关联。内部关系也可以采用下面语句查询，效果是一样的：

```
select course.course_name as 课程名,course_address as 上课地点,course_time as 上课时间,
teacher.teacher_name as 任课教师 from course inner join teacher on course.teacher_id =
```

teacher.teacher_id

将 inner join 改换成 left join 或 right join,上面的 select 语句就变成左关联或右关联查询了。

(2) 查询学生的成绩,显示"学号""姓名""课程名"和"成绩"。

查询显示的 4 个字段来自 3 个表:"学号"和"姓名"可来自 student 表中的 student_no 和 student_name,"课程名"来自 course 表中的 course_name,"成绩"来自 score 表中的 score。student 表与 score 表中的 student_no 相关联,course 表与 score 表通过 course_no 关联。查询的语句为:

select student.student_no as 学号,student_name as 姓名,course.course_name as 课程名,score.score as 成绩 from student,course,score where student.student_no = score.student_no and course.course_no = score.course_no

同样,该语句也可以写为:

select student.student_no as 学号,student_name as 姓名,course.course_name as 课程名,score.score as 成绩 from student inner join score on student.student_no = score.student_no inner join course on course.course_no = score.course_no

2.7.5 增加、删除和更新记录

1. 增加一条新记录

insert into teacher(teacher_id,teacher_name,teacher_gender,teacher_phone) values('t4','刘甜甜','女','010-6875357')

由于 id 字段设置的是自动增加,故不能通过代码为该字段赋值。
增加单条记录的 insert 语句的格式如下:

insert into 表名(字段 1,字段 2) values(值 1,值 2)

2. 批量增加记录

insert into 表名 1(字段 1,字段 2) select 字段 1,字段 2 from 表名 2 where 条件

将表名 2 中符合条件的字段 1、字段 2 的值,批量增加到表名 1 对应的字段中。使用时要注意表名 2 的字段必须与表名 1 的相一致;否则出现错误。

3. 删除记录

delete from 表名 where 删除条件

例如:

delete from student where student_no = '41321059'

SQL 语句可以相互嵌套。如果要删除 student 表中那些在 score 表中 score<60 的记录:delete from student where student_no in(select student_no from score where score<60),语句执行时,首先通过 select student_no from score where score<60,从 score 表中找

到不及格学生的 student_no,然后从 student 表中删除这些 student_no 的记录。

4. 更新记录

update 表名 set 字段 1 = 值 1,字段 2 = 值 2 where 更新条件

例如：

update score set score = score + 5 where score < 60

为不及格的同学加 5 分,如果要为女生加 5 分,需要使用 SQL 嵌套：

update score set score = score + 5 where student_no in(select student_no from student where gender = '女')

2.7.6 操作表结构的 SQL

某些情况下,表的字段不是预先固定好的,需要通过程序动态生成和修改。

1. 建立一个新表

create table 工资(雇员号 char(8) not null, Primary Key(雇员号),工资日期 date,基本工资 decimal(8,2),岗位津贴 decimal(8,2),id int IDENTITY(1,1) NOT NULL)

使用 create 创建"工资"表,表的主键是"雇员号",值不能为空；id 整型,从 1 开始自动编号。

2. 修改表结构

增加字段：alter table 表名 add 字段名 数据类型

删除字段：alter table 表名 drop 字段名

给已经建立的"工资"表增加一列"所得税"：

alter table 工资 add 所得税 decimal(6,2)

3. 删除数据表

drop table 表名

将"工资"表删除：

drop table 工资

习题与思考

(1) 在 student 表、score 表、course 表、teacher 表中输入记录,分别见图 2-17～图 2-20。

(2) 将 student 表中最后一条记录的 student_no 由 42405002 改为 41405002,查看 score 表中 student_no 是 42405002 的记录如何发生变化？

(3) 删除 student 表中 student_no 等于 41405002 的记录,查看 score 表中 student_no 等于 41405002 的记录是否被自动删除？

	student_no	student_name	gender	email	identification	id
1	30012035	楚云飞	男	zsf@sohu.com	230304200611201537	1
2	40012030	赵三	男	zs@sohu.com	230304200010302378	3
3	41321059	李孝诚	男	lxc@163.com	110106199802030013	4
4	41340136	马小玉	女	mxy@126.com	130304199705113028	5
5	41355062	王长林	男	wcl@163.com	110106199512035012	6
6	41361045	李将寿	男	ljc@126.com	14510619901203503X	7
7	41361258	刘登山	男	lds@126.com	130030520001234235	8
8	41361260	张文丽	女	zhangwl@163.com	140305200106202348	9
9	41401007	鲁宇星	男	lyx@sohu.com	110108200012035012	10
10	41405002	王小月	女	wxy@263.com	130304199901115047	11

图 2-17　student 表中内容

41321059	k01	89.0	1
41321059	k02	90.0	2
41321059	k03	70.0	3
41340136	k01	50.0	4
41340136	k02	80.0	5
41340136	k03	77.0	6
41340136	k04	60.0	7
41355062	k01	90.0	8
41355062	k02	88.0	9
41355062	k03	75.0	10
41361045	k01	66.0	11
41361045	k02	85.0	13
41361045	k03	90.0	14
41401007	k01	82.0	15
41401007	k02	62.0	16
41401007	k03	74.0	17
41405002	k01	90.0	21
41405002	k02	84.0	22
41405002	k03	85.0	23
41405003	k01	81.0	18
41405003	k02	72.0	19
41405003	k03	76.0	20

图 2-18　score 表中内容

course_no	course_name	course_address	course_time	teacher_…	id
k01	ASP网络数据库	教学楼107	周二8:00-10:00	t1	1
k02	高等数学	主楼203	周一8:00-10:00	t2	2
k03	英语	逸夫楼305	周四10:00-12:00	t3	3
k04	人工智能	主楼305	周三8:00-10:00	t1	4

图 2-19　course 表中内容

teacher_ID	teacher_name	teacher_gender	teacher_phone	id
t1	刘小明	男	82315647	1
t2	赵玉琴	女	62335489	2
t3	肖宝强	男	62314356	3

图 2-20　teacher 表中内容

(4) 将图 2-20 中最后一条记录的 teacher_id 由 t1 改为 t4 是否可以？为什么？

(5) student 表与 score 表的关联线的属性"强制外键约束"设置为"是"，能否直接在 score 表中插入表 2-5 中记录（第 1 行是字段名）？

表 2-5 要求插入的记录

student_no	course_no	score
52405001	K01	50

（6）在 SQL Server 中写出查询学生成绩的 select 语句，各表间采用内关联的方式，显示 student_no、student_name、course_name、score。

（7）查询某老师所开设的课程，显示字段如下：

teacher_ID, teacher_name, course_name, course_address, course_time

第 3 章

网页编程基础知识

(1) HTML、CSS。
(2) JavaScript 基本语法。
(3) JavaScript 中的对象。
(4) JavaScript 解析 JSON。
(5) XML 和 Ajax 的基本概念。

服务器全天候地响应来自客户端不断发出调用网页、图像、声音、视频等请求,服务器得到客户端的请求后,从服务器上找到相应的资源,然后以标准的 HTML 形式返回给客户端,客户端上的浏览器负责解释服务器返回的 HTML。

HTML 是一种描述网页的语言,其不属于编程语言,是一种标记语言。标记语言是一种用来对文本进行注释的语言,其提供的注释与原始文本在语法上是可区别的,也称置标语言。

HTML 5 是 HTML 最新的修订版本,2014 年 10 月由万维网联盟(W3C)完成标准制定,设计目的是为了在移动设备上支持多媒体。

3.1 HTML

3.1.1 HTML 文档结构

HTML 文档是由 HTML 标记按照一定的结构定义的。在 VS 中建立的 HTML 文件,自动生成的结构如下:

```
<!DOCTYPE html>
<html xmlns="http://www.w3.org/1999/xhtml">
<head>
<meta http-equiv="Content-Type" content="text/html; charset=utf-8"/>
<title></title>
```

```
</head>
<body>
</body>
</html>
```

说明：

① 大多数 HTML 标记成对出现，如<head></head>。

② 第1行，声明为 HTML5 文档。

③ 第2行<html>是 HTML 页面的根标记，加上 xmlns 属性，定义了一个命名空间，浏览器使用该命名空间解析文档中所有的标记。

④ 第 3 行＜head＞，定义文档的头部，常常包含＜title＞＜meta＞＜style＞＜script＞等。

⑤ 第 4 行＜meta＞是 HTML 的元标记，其中包含了对应 HTML 的信息。其属性 http-equiv 类似于 HTTP 的头部协议，它回应给浏览器一些有用的信息，以帮助其正确和精确地显示网页内容；content 为内容类型；charset 为编码类型，网页中有中文时，编码类型要设置为 utf-8，否则网页会出现乱码。

例 3-1 建立一个 HTML 文件，浏览器打开后效果如图 3-1 所示。

图 3-1　例 3-1 在 Google Chrome 浏览器中显示效果

例 3-1 中 HTML 的文档内容如下：

```
<!DOCTYPE html>
<html xmlns = "http://www.w3.org/1999/xhtml">
<head>
<meta http-equiv = "Content-Type" content = "text/html; charset = utf-8"/>
    <title>Introduce of jQuery</title>           Ⓐ
    <style>
        .bookImage
        {
            width:290px;
            height:190px;
        }
        .bookIntro
```

```
            {
                height:100px;
                width:290px;
                text-align:justify;
                border:solid #000 1px;
                display:block;
            }
            .main
            {
                text-align:center;
                margin:0 auto;
                height:100px;
                width:290px;
            }
        </style>
    </head>
    <body>
    <div class="main">
            <h2>一本很不错的 jQuery 书籍</h2>                                         Ⓑ
            <img src="jQuery.jpg" alt="jQuery" class="bookImage"/>                  Ⓒ
             <div class="bookIntro">
                    第 1 章:在掌握了 HTML 和 CSS 后,希望再增加一项编写脚本的技能,学习使用 jQuery,就能
            够动态地修改 Web 页面!                                                    Ⓓ
            </div>
            <div>
              <a href="www.cepp.sgcc.com.cn">请到××出版社购买</a>                    Ⓔ
            </div>
    </div>
    </body>
</html>
```

说明:

① 文档中代码ⒶⒷⒸⒹⒺ显示为图 3-1 中的ⒶⒷⒸⒹⒺ。

② 在<head></head>中的<style></stlye>定义的是 CSS,用于控制标记的显示格式,具体用法见 CSS 相关内容。

③ <img src="url" alt="可替换的文本内容">通过 src 属性指定图像文件所在的位置,定义网页中的图像。在浏览器无法载入图像时,alt 属性用来为图像定义一串预备的可替换的文本。

④ <a href="URL">链接文本</a>HTML 的超链接,当单击"链接文本"时,网页转到 URL 指定的页面。

3.1.2　HTML 表单

HTML 表单用于搜集不同类型的用户输入,表单中可用于输入的表单元素有文本框、下拉列表框、单选按钮、复选框等。表单使用<form>设置,例如:

```
<form id="ff" method="get/post" action="服务器端程序名">
…
```

各种<input>元素
</form>

form 中常用的属性有两个：method 和 action。method 提交表单的方法有 get 和 post 两种；action 提交表单的处理程序。当用 Ajax 提交表单时，不需要设置 action。

1. 文本框

文本框通过<input type="text">标记来设定，当用户要在表单中输入字母、数字等内容时，就会用到文本框。

```
<form id="ff">
    姓名:<input type="text" name="sname"/><br/>
    年龄:<input type="text" name="age"/>
</form>
```

浏览器显示结果为：

2. 密码字段

密码字段通过标记<input type="password">来定义：

```
<form id="ff">
    姓名:<input type="text" name="sname"/><br/>
    年龄:<input type="text" name="age"/><br/>
    密码:<input type="password" name="pwd"/><br/>
</form>
```

浏览器显示结果为：

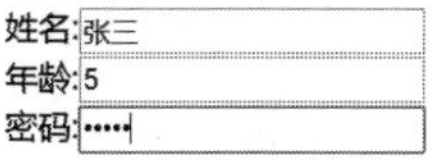

3. 单选按钮

<input type="radio">标记定义了表单单选按钮选项：

```
<form>
    <input type="radio" name="sex" value="male">男<br/>
    <input type="radio" name="sex" value="female">女
</form>
```

浏览器显示结果为：

⊙男
⊙女

4. 复选框

<input type="checkbox">定义了复选框。用户需要从若干给定的选择中选取一个或若干选项：

```
<form>
```

```
        what's your hobby?< br/>
        < input type = "checkbox" name = "hobby" value = "Bike"> music < br/>
        < input type = "checkbox" name = "hobby" value = "Car"> swim < br/>
        < input type = "checkbox" name = "hobby" value = "Car"> football
</form>
```

浏览器显示结果为：

```
what's your hobby?
☐music
☐swim
☐football
```

5. 提交按钮

```
< form name = "input" action = "html_form_action.ashx" method = "get">
        用户名：< input type = "text" name = "user">< br/>
        < input type = "submit" value = "提交">
</form>
```

<input type="submit">定义了提交按钮。当用户单击提交按钮时，表单的内容会被传送到由<form action="">指定的另一个文件，由定义的这个文件通常会对接收到的输入数据进行相关的处理。浏览器显示结果为：

HTML 5 为表单提供了新的输入元素，如 color、date、email 等。要注意，并非所有浏览器都支持 HTML 5，不支持时，这些输入会显示为常规的输入框。

6. color

color 类型用在 input 字段主要用于选取颜色。

```
< form>
        选择你喜欢的颜色：< input type = "color" name = "favcolor">
        < input type = "submit" value = "提交">
</form>
```

浏览器显示结果为：

选择你喜欢的颜色：▇ 提交

7. date

date 类型允许你从一个日期选择器选择一个日期。

```
< form>
        选择你的生日：< input type = "date" name = "bday">
</form>
```

在浏览器中选择日期，出现效果如下：

8. email

email 类型用于应该包含 email 地址的输入域,在提交表单时,会自动验证 email 域的值是否合法有效。

```
<form>
E-mail:<input type="email" name="email">
    <input type="submit">
</form>
```

在浏览器中,如果输入的 email 不符合要求,出现效果如下:

9. number

number 类型用于应该包含数值的输入域,可设定所接受的数字范围。

```
<form>
数量(1 到 10 之间):<input type="number" name="quantity" min="1" max="10">
    <input type="submit">
</form>
```

在浏览器中,如果输入的值不在指定范围,则出现效果如下:

3.1.3 表格

表格的语法结构示例如下:

```
<table>
    <tr><td>Month</td><td>saving</td></tr>
    <tr><td>January</td><td>$100</td></tr>
    <tr><td>February</td><td>$80</td></tr>
    <tr><td>Sum</td><td>$180</td></tr>
</table>
```

有的表格中使用<thead><tbody><tfoot>分别定义表格的表头、表格主体、表格脚注,从而使浏览器有能力支持独立于表格表头和表格页脚的主体滚动。当打印包含多个页

面的长表格时,表格的表头和页脚可被打印在包含表格数据的每张页面上。此时表格的结构变为:

```
<table>
    <thead><tr><td>Month</td><td>saving</td></tr></thead>
    <tbody><tr><td>January</td><td>$100</td></tr>
    <tr><td>February</td><td>$80</td></tr></tbody>
    <tfoot><tr><td>Sum</td><td>$180</td></tr></tfoot>
</table>
```

3.2 CSS

CSS用于控制网页的样式和布局,最新的CSS标准是CSS3。

从例3-1中可看出,HTML中使用CSS可控制页面元素的外观和位置,页面中使用CSS,以解决页面内容和格式分离;使用样式单独存放为一个扩展名为CSS的文件,可以使多个页面共享样式。样式一般使用<style></style>定义在<head></head>中。CSS要控制页面中元素,需要掌握CSS选择元素的方式,即CSS选择器的使用方法。

1. CSS选择器

```
                这是一个元素选择器,它会选择HTML文档中的所有h1元素
h1{
    text-align:center;
}

        CSS类选择器,从一个点开始。利用类选择符可以选择一
        组元素。类名第1个字符不能使用数字
.my_class{                  设置属性值的格式为属性:属性值;
    position:absolute;
    border:solid 1px;
}

        CSS ID总是以#开始。如果想要选择一个且仅一
        个元素,就要使用CSS ID
#my_id{
    color:#3300FF;
}
```

例3-2 使用DIV+CSS页面布局,在浏览器中运行后,页面自动居中,页面距离浏览器顶部的距离是0,效果如图3-2所示。

HTML的文档内容如下:

```
1   <!DOCTYPE html>
2   <html xmlns="http://www.w3.org/1999/xhtml">
3   <head>
4   <meta http-equiv="Content-Type" content="text/html; charset=utf-8"/>
5   <title>my courses</title>
6   <style>
7   *
8   {
```

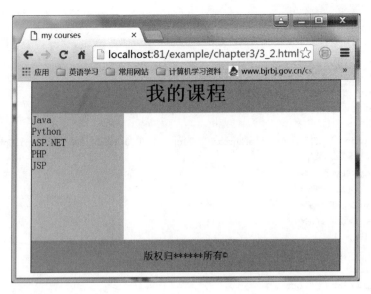

图 3-2　CSS+DIV 布局页面

```
9      margin: 0px;
10     padding: 0px;
11  }
12  #header
13  {
14     background-color:#FFA500;
15     height:50px;
16     text-align:center;
17  }
18  #container
19  {
20     width:500px;
21     margin:0 auto;
22     border:solid 1px;
23  }
24  .content
25  {
26     float:left;
27     width:300px;
28     height:50px;
29     position:relative;
30     left:6px;
31     top:0px;
32  }
33  .bottom
34  {
35     background-color:#FFA500;
36     text-align:center;
37     clear:left;
38     height:50px;
39     line-height:50px;
```

```
40  }
41  </style>
42  </head>
43  <body>
44  <div id = "container">
45      <div id = "header">
46          <h1 style = "margin-bottom:0 auto">我的课程</h1>
47      </div>
48      <div id = "left" style = "background-color:#FFD700;height:200px;width:150px;float:left;">
49          <p>Java</p>
50          <p>Python</p>
51          <p>ASP.NET</p>
52          <p>PHP</p>
53          <p>JSP</p>
54      </div>
55      <div class = "content" id = "intro">
56      </div>
57      <div class = "bottom">
58          版权归******所有 &copy;
59      </div>
60  </div>
61  </body>
62  </html>
```

说明：

① 第6～41行定义了CSS，CSS设置属性的格式是：

属性:属性值

② 要理解CSS中的布局，在CSS中有一个盒子模型，模型中所有HTML元素都可以看作盒子，图3-3是盒子模型示意图。

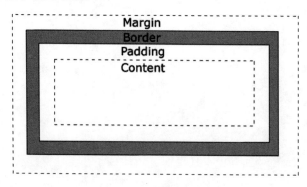

图3-3 盒子模型示意图

③ 第7行 * 表示所有元素；margin:0px 表示所有元素的外边距是0；margin:5px 10px 表示上外边距和下外边距是5px，左外边距和右外边距10px；margin:0 auto 中 auto 表示左右外边距根据宽度自适应相同值。margin:10px 5px 15px 20px 表示上外边距是10px，右外边距是5px，下外边距是15px，左外边距是20px。

margin(外边距)：清除边框外的区域，外边距是透明的；border(边距)：围绕在内边距和内容外的边框;pagging(内边距):清除内容周围的区域,内边距是透明的;Content(内容):盒子的内容,显示文本和图像等。

④ 第12行♯header使用的是CSS ID选择器定义的方式,第45行中要通过设置<div>的id属性使用该样式。

⑤ 第14行background-color设置元素的背景色,♯FFA500是CSS中用16位数表示颜色的一种方法,FF表示红色,A5表示绿色,00表示蓝色,♯FF0000表示纯红色。也可以用RGB(R,G,B)函数表示颜色,R、G、B取值0～255,分别表示红、绿、蓝,或者使用RGB百分比值(R%,G%,B%)来表示颜色,如RGB(100%,0%,0%)表示红色。

⑥ 第15行height:50px表示高度是50像素。CSS中长度、高度的单位的px(像素)、pt(1pt等于1/72英寸)、cm(厘米)、mm(毫米)、in(英寸)、em(1em等于当前的字体尺寸)、ex(一个ex是一个字体的x-height,x-height通常是字体尺寸的一半)、%(百分比)。用得最多的是px。

⑦ 第24行.content是CSS类选择器,第55行要通过设置<div>的class="content"使用该样式。

⑧ 第26行float定义元素在哪个方向上浮动。CSS中任何元素都可以浮动;第29行position属性,设置元素的定位信息,取值可以是abosolute、relative等,设置position:relative后,left:6px;top:0px;是相对于第48行的<div>元素而言。

⑨ 第37行clear属性规定元素的那一侧不允许其他浮动元素,取值有left、right、both。

⑩ 第39行line-height设置行间的距离,设置该属性的目的是使第58行文字垂直居中。

2. CSS的创建

插入样式表的方法有外部样式表、内部样式表、内联样式。

(1) 外部样式表。当有许多页面需要使用相同的样式时,可以将这些样式保存在一个单独的扩展名为css的文件中,如mycss.css中,该文件可以用记事本书写,也可以在VS中,通过"资源管理器"→"添加"→"样式表"命令,建立自己的css文件。下面是一个mycss.css文件的示例:

```
body {
    margin:0 auto;
    padding:0;
}
p
{
    text-indent:4px;
}
♯header
{
  background-color:♯FFA500;
  height:50px;
  text-align:center;
}
```

在每个需要使用这些样式的页面的<head></head>中使用<link>标记链接到样式表。例如:

```
<head><link rel = "stylesheet" type = "text/css" href = "mycss.css"></head>
```

(2) 内部样式表。当单个文档需要特殊的样式时，就应该使用内部样式表。在页面的 `<head><style></style><head>` 定义样式，然后在 HTML 开始的元素中通过设置属性 ID 或者 class，使用定义好的样式。

(3) 内联样式。仅当在个别 HTML 元素上使用样式时采用这种方法。使用内联样式，需要在相关的标记内使用样式(style)属性。style 属性可以包含任何 CSS 属性。例 3-2 第 48 行使用的就是内联样式。内联样式将格式的表现和内容混杂在一起，会失掉样式表的许多优势。

3.3 通过 JavaScript 为网页增加动作

例 3-3 使用 HTML+CSS 可以很好地表现网页的内容和格式，现在需要为例 3-2 中的页面增加动作：希望当单击 Java、Python 等书名时，在网页右侧空白处显示出关于该书的内容介绍。

在例 3-2 的 `<head></head>` 中增加以下代码：

```
<script>
    function GoToURL(introduce)
    {
      document.getElementById("intro").innerHTML = introduce;
    }
</script>
```

将例 3-2 中第 48～54 行替换为以下内容：

```
1  <div id = "left" style = "background - color: #FFD700;height:200px;width:150px;float:
   left;">
2    <div><a href = "#" onclick = "GoToURL('这是一门介绍 Java 的好课程')">Java</a></div>
3    <div><a href = "#" onclick = "GoToURL('这是一门介绍 Python 的好课程')">Python</a></div>
4    <div><a href = "#" onclick = "GoToURL('这是一门介绍 ASP.NET 的好课程')">ASP.NET</a>
     </div>
5    <div><a href = "#" onclick = "GoToURL('这是一门介绍 PHP 的好课程')">PHP</a></div>
6    <div><a href = "#" onclick = "GoToURL('这是一门介绍 JSP 的好课程')">JSP</a></div>
7  </div>
```

说明：

① JavaScript 程序代码一般放在 `<script></script>` 中。`<script>` 可以放在 `<head></head>` 间，也可以放在 `<body></body>` 的任意位置。

② JavaScript 定义函数语法为 function 函数名([形参]){函数体}，如果函数有返回值，使用 return 返回值；如果函数需要在多个网页中共享，与 CSS 外部链接类似，可以将 JavaScript 程序代码放在一个扩展名为 .js 的文件中，如 myJS.js，该文件中不能包含 `<script>`，内容类似于

```
function GoToURL(introduce)
{
```

```
        document.getElementById("intro").innerHTML = introduce;
}
function GetByID()
{
  …;
}
```

在需要使用这些函数的页面的＜head＞或者＜body＞中加入:

`<script src = "myJS.js"></script>`

当然在引用myJS.js文件时一定要注意该文件与当前网页的位置关系,上面这种写法表示myJS.js文件与当前网页文件在同一文件夹下。

也可以直接在HTML的开始元素中添加JavaScript。例如:

`<button onclick = "window.alert('hello')">单击我</button>`

显然,这种方式下编写的代码不宜过长。

③ JavaScript控制浏览器,需要使用一个DOM(Document Objects Models,文档对象模型),在该模型中定义了window、document等对象,各对象都有相应的属性和事件。Document可理解为当前浏览的页面,document.getElementById根据页面上元素的id(页面上元素的id是唯一的)找到该元素,页面元素有一个方法innerHTML,用于设置(获得)该元素间的HTML。

JavaScript获得页面上元素的方法还有getElementsByTagName()、getElementsByName(),由于页面上同一标记可以出现多次,相同名字的标记也可以有多个,故getElementsByTagName("p")[0]是页面上出现的第一个p元素,getElementsByName("favorite")[0]是页面上第一个name为favorite的元素。

DOM一直帮助HTML、CSS和JavaScript有效地协同工作,它提供了一个标准化框架,现在大部分浏览器都支持这一框架。DOM就像是一棵树,有一个根,还有分支,分支的末端就是节点,DOM示意图见图3-4。

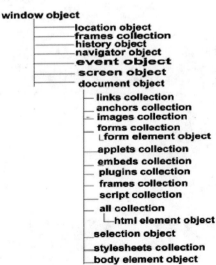

图3-4 DOM示意图

为了降低 JavaScript 通过 DOM 模型操作页面各元素的复杂性，在 JavaScript 基础上，产生了各种 JavaScript 的库，如本书中介绍的 jQuery，不仅降低 UI(User Interface，用户界面，或者称前台)编程的难度，也增强了浏览器的兼容性。基于这些库做前台程序开发，可降低开发的难度，使界面更加丰富和友好。

④ 在 HTML 开始的元素上增加事件，所有事件都是以 on 开关的，如 onclick、onchange、onfocus、onblur、onmouseover、onmouseout 等，第 2～6 行在＜a href＝"＃" onclick＝"GoToURL()"＞设置为空的超链接，当单击该链接时，调用＜script＞＜/script＞中定义好的函数，将要显示的内容作为实参传递给该函数的形参 introduce。

浏览器中打开该文件，运行效果如图 3-5 所示，单击某链接，如 ASP.NET，图 3-5 变为图 3-6。

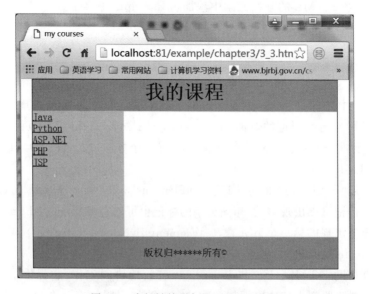

图 3-5 为超链接增加 JavaScript 代码

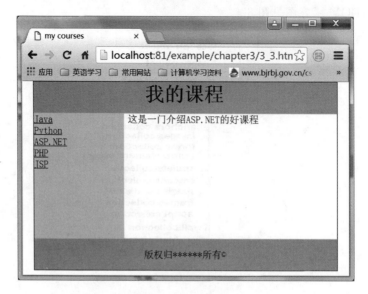

图 3-6 单击 ASP.NET 超链接时的显示效果

说明：图3-5中各课程名是一个链接，默认情况下该链接的文字下面有一下划线，如果想要取消下划线，可增加下列样式：

```
a:link
{
    text-decoration:none;
}
```

上面示例中，页面左侧的课程名是在网页中直接输入，如果课程名和课程内容发生改变，需要修改网页中的HTML，这种静态网页编程的方式显然满足不了人们的要求。在静态网页的基础上，需要掌握动态网页编程技术，课程名和课程介绍都从数据库中调取。

3.4 用JavaScript修改HTML元素的样式

JavaScript利用DOM，可以修改页面元素的显示样式（图3-7），修改的方法有以下几种。

图3-7 改变背景色

1. document.getElementById("id").style.property="值"

```
<body>
<div id="main">
    One morning, when Gregor Samsa woke from troubled dreams.
</div>
<button onclick="changeStyle()">改变背景</button>
<script>
    function changeStyle() {
        var obj = document.getElementById("main");
        obj.style.backgroundColor = "green";
    }
</script>
</body>
```

2. document.getElementById("id").style.cssText="css 属性:值"

将上面的脚本改为：

```
<script>
    function changeStyle() {
        var obj = document.getElementById("main");
        obj.style.cssText = "background-color:green";
    }
</script>
```

要注意 cssText 中设置的属性名与 style. 属性名并不完全一致，如 backgroundColor 与 background-color。cssText 设置的属性名采用的是 CSS 中的属性，而 obj. style 中采用的属性是 HTML 中的属性，要注意区分。

3. document. getElementById("id"). setAttribute("class", "style")

```
<!DOCTYPE html>
<html xmlns = "http://www.w3.org/1999/xhtml">
<head>
<meta http-equiv = "Content-Type" content = "text/html; charset = utf-8"/>
    <title></title>
</head>
    <style>
        .backSet
        {
            background-color:blue;
            color:white;
        }
    </style>
<body>
<div id = "main">
    One morning, when Gregor Samsa woke from troubled dreams.
</div>
<button onclick = "changeStyle()">改变背景</button>
<script>
    function changeStyle() {
        var obj = document.getElementById("main");
        //obj.className = "backSet";
        obj.setAttribute("class","backSet");
    }
</script>
</body>
</html>
```

说明：先定义好类样式 backSet，然后使用 obj. className = "backSet" 或者 obj. setAttribute("class","backSet");修改元素的样式。

3.5 JavaScript 编写简单的扑克游戏

例 3-4 有 52 张扑克牌（没有大小王），按照花色分别放在 cards 下面的 Clubs、Diamonds、Hearts、Spades 文件夹下，每个文件夹下图片的文件名都相同，但花色不一样。图 3-8 是 Clubs 下的扑克牌及文件名。

要求单击一叠扑克的图片，一次发出 5 张不同的牌，自动将牌上点数相加（每张 J、Q、K 的点数是 0.5）。运行效果如图 3-9 所示。52 张扑克牌放在 4 个文件夹中，编程时如何存储这些扑克牌的信息？使用数组是解决问题的一个好方法。

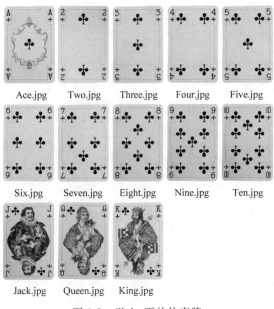

图 3-8　Clubs 下的扑克牌

图 3-9　运行结果

3.5.1　数组对象

1. 数组的定义

存储数据最简单的方法是使用变量，如 var x＝50；将 50 存储在变量 x 中，并为变量赋值，JavaScript 是一种弱类型语言，变量声明时使用 var 即可（也可以不用）。数组提供了一种更好的存储方式，可以使用一个变量名赋多个值：

使用new关键字创建一个空的数组
var arr1=new Array();

使用new关键字创建一个数组，并且对数组赋初值
var arr2=new Array("China", "USA", "Australia", "France");

没有使用new关键字创建一个数组，下面是使用中括号[ ]直接为数组赋值
var arr3=["China", "USA", "Australia", "France"];

上述3种建立数组的方法没有任何区别。

2. 数组中元素的访问

3. 数组中元素的增加和更新

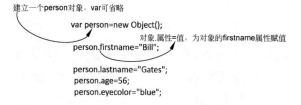

数组是对象，最常用的属性是 length，表示数组中有多少个元素。可以使用"数组名.length"来访问 length 的属性。

3.5.2 自定义对象

当需要存储某个事物的多个变量时，可以使用对象来存储数据。描述这些对象的变量，称为属性。对象中可以包含函数，这些函数称为方法。

下面定义一个 person 对象。

1. 创建直接的方法

```
建立一个person对象，var可省略
    var person=new Object();
                    对象.属性=值，为对象的firstname属性赋值
    person.firstname="Bill";
    person.lastname="Gates";
    person.age=56;
    person.eyecolor="blue";
```

或者

```
var person = {firstname:"Bill",lastname:"gates",age:56,eyecolor:"blue"};
```

这种方法构建的对象是一次性的，只是使用对象存储了多个属性。

2. 使用对象构造函数构建可重用的对象

```
person是对象名         属性的4个参数
    function person(firstname,lastname,age,eyecolor)
    {
        this.firstname=firstname;   firstname是对象属性
                                    通过参数firstname设置
                                    对象的属性值
        this.lastname=lastname;
        this.age=age;
        this.eyecolor=eyecolor;
    }
```

上面属性的4个参数值正好与要设置的4个属性相同,如果属性的4个参数改为a、b、c、d,则对象的构造函数改为:

```
function person(a,b,c,d)
{
    this.firstname = a;
    this.lastname = b;
    this.age = c;
    this.eyecolor = d;
}
```

由于a、b、c、d意义不明确,使程序可读性变差。

构造函数中也可以定义函数,称之为方法。例如:

```
function person(firstname,lastname,age,eyecolor)
{
    this.firstname=firstname;
    this.lastname=lastname;
    this.age=age;
    this.eyecolor=eyecolor;
    this.getName = function ()      定义一个方法getName
    {
        return this.firstname + " " + this.lastname;
    };
}
```

构造函数建好后,可以定义许多对象的新实例。例如:

```
var myFather = new person("Bill","Gates",56,"blue");
var myMother = new person("Steve","Jobs",48,"green");
```

下面访问这两个对象的属性和方法:

```
var myFather=new person("Bill","Gates",56,"blue");
                                        输出结果: Gates
alert(myFather.lastname);
                                        getName后面要有(),输出结果: Bill Gates
alert(myFather.getName());
var myMother=new person("Steve","Jobs",48,"green");
                                        输出结果: Jobs
alert(myMother.lastname);
                                        输出结果: Steve Jobs
alert(myMother.getName());
```

3.5.3 扑克牌中的页面

建立一个 HTML 页面和 CSS 文件,如 3_4.html 和 3_4.CSS。
3_4.html 内容如下:

```html
<!DOCTYPE html>
<html xmlns="http://www.w3.org/1999/xhtml">
<head>
<meta http-equiv="Content-Type" content="text/html; charset=utf-8"/>
    <title></title>
    <link href="3_4.css" rel="stylesheet"/>
</head>
<body>
    <div id="main">
        <h1>单击扑克图片,每次随机发5张不同的牌</h1>
        <h3 id="totalScore"></h3>
        <div id="getCards">
        </div>
        <div id="controls">
        <img id="dealImage" src="joker.jpg" onclick="deal()" alt="随机发牌" title="随机发牌"/>
        </div>
    </div>
     <script type="text/javascript" src="card.js"></script>
</body>
</html>
```

3_4.css 内容如下:

```css
#getCards
{
    border: 1px solid gray;
    height: 250px;
    width: 835px;
}
h3
{
    display: inline;
    padding-right: 40px;
}
#dealImage
{
    width:100px;
    height:100px;
}
```

3.5.4 扑克牌中的属性和方法

例 3-4 中的扑克牌需要什么属性?
每个花色的牌表现为不同文件名的图片,需要一个描述图片文件名的属性 name;4 种花色的扑克牌存放在 4 个文件夹中,需要一个描述文件夹的属性 suit;需要一个描述扑克点数的属性 value,这样扑克对象 card 需要 name、suit、value 这 3 个属性来描述。

新建立一个 card.js 文件,输入以下代码(不输入表示代码行的数字):

```
var usedCards = new Array();
var TotalScore = 0;     //存储点数
function card(name, suit, value) {
    this.name = name;
    this.suit = suit;
    this.value = value;
}
var deck = {
new card('Ace', 'Hearts', 1),
new card('Two', 'Hearts', 2),
new card('Three', 'Hearts', 3),
new card('Four', 'Hearts', 4),
new card('Five', 'Hearts', 5),
new card('Six', 'Hearts', 6),
new card('Seven', 'Hearts', 7),
new card('Eight', 'Hearts', 8),
new card('Nine', 'Hearts', 9),
new card('Ten', 'Hearts', 10),
new card('Jack', 'Hearts', 0.5),
new card('Queen', 'Hearts', 0.5),
new card('King', 'Hearts', 0.5),
new card('Ace', 'Diamonds', 1),
new card('Two', 'Diamonds', 2),
new card('Three', 'Diamonds', 3),
new card('Four', 'Diamonds', 4),
new card('Five', 'Diamonds', 5),
new card('Six', 'Diamonds', 6),
new card('Seven', 'Diamonds', 7),
new card('Eight', 'Diamonds', 8),
new card('Nine', 'Diamonds', 9),
new card('Ten', 'Diamonds', 10),
new card('Jack', 'Diamonds', 0.5),
new card('Queen', 'Diamonds', 0.5),
new card('King', 'Diamonds', 0.5),
new card('Ace', 'Clubs', 1),
new card('Two', 'Clubs', 2),
new card('Three', 'Clubs', 3),
new card('Four', 'Clubs', 4),
new card('Five', 'Clubs', 5),
new card('Six', 'Clubs', 6),
new card('Seven', 'Clubs', 7),
new card('Eight', 'Clubs', 8),
new card('Nine', 'Clubs', 9),
new card('Ten', 'Clubs', 10),
new card('Jack', 'Clubs', 0.5),
new card('Queen', 'Clubs', 0.5),
```

```
47    new card('King', 'Clubs', 0.5),
48    new card('Ace', 'Spades', 1),
49    new card('Two', 'Spades', 2),
50    new card('Three', 'Spades', 3),
51    new card('Four', 'Spades', 4),
52    new card('Five', 'Spades', 5),
53    new card('Six', 'Spades', 6),
54    new card('Seven', 'Spades', 7),
55    new card('Eight', 'Spades', 8),
56    new card('Nine', 'Spades', 9),
57    new card('Ten', 'Spades', 10),
58    new card('Jack', 'Spades', 0.5),
59    new card('Queen', 'Spades', 0.5),
60    new card('King', 'Spades', 0.5)};
```

说明：

① 第 1 行定义数组 usedCards，用于存储已经随机抽出的扑克。

② 第 8~60 行定义了数组 deck，数组中的元素是 52 个 card 对象构成的，每个 card 对象依次传入 name、suit、value 第 3 个参数。

随机抽取扑克牌，需要的方法有以下几个。

(1) getRandom()：产生 1~52 个随机整数的方法。

```
function getRandom(num) {
    var my_num = Math.floor(Math.random() * num);
    return my_num;
}
```

说明：JavaScript 的对象有 3 类，即浏览器对象、JavaScript 内部对象和自定义对象。浏览器对象是文档对象模型（DOM）中规定的对象，如 HTML 元素对象、文档对象、窗口对象等；内部对象包括 JavaScript 常用的对象，如数组、日期、字符串、数学对象等。Math 是 JavaScript 的数学对象，Math.floor()返回小于或等于一个给定数字的最大整数，Math.random()可返回介于 0~1 的一个随机数，Math.floor(Math.random() * num)返回 0~num 间一个随机整数。

(2) hit()：抽取牌的方法。

```
1   function hit()
2   {
3       var validCards = true;
4       while (validCards == true) {
5           var index = getRandom(52);
6           var c = deck[index];
7           if (usedCards[index] != c)           //抽过的牌不能再抽取
8           {
9               usedCards[index] = c;
10              TotalScore += c.value;
11              document.getElementById("totalScore").innerHTML = "牌的总点数：" + TotalScore;
12              document.getElementById("getCards").innerHTML += "<img src='" + "cards/" + c.suit + "/" + c.name + ".jpg'>";
```

```
13              validCards = false;
14          }
15      }
16  }
```

说明：

① 第 4～15 行循环,直到抽出没有抽取过的牌。用 validCards 变量控制循环的开始和结束。

② 第 7～14 行通过判断,保证抽过的牌不再被抽中。

③ 第 6 行从 deck 数组中取得索引是 index 的那张扑克,如 index=2,得到的是 card('Two', 'Hearts', 2)。

④ 第 9 行将抽取出的牌存放到索引号是 index 的数组中。

⑤ 第 12 行通过拼接字符串,输出抽中的扑克牌。c.suit 是扑克牌的文件夹名,c.name 是扑克的图片文件名。

(3) deal()：循环抽取 5 次牌。

```
function deal()
{
    reset();
    for (var i = 0; i < 5; i++)
    {
        hit();
    }
}
```

说明：JavaScript 的循环语句 for(var i=0;i<5;i++){}与 C♯的相似,i 是循环变量,循环的起始值是 0,每次增加 1,直到 i>=5 时退出循环。

(4) reset()：重置变量到抽取前的状态。

```
function reset()
{
    document.getElementById("getCards").innerHTML = "";
    document.getElementById("totalScore").innerHTML = "";
    usedCards.length = 0;            //清空数组
    TotalScore = 0;
}
```

3.6 window 对象控制定时效果

每次访问者在浏览器中打开一个新窗口时,就会创建一个 window 对象,window 对象是 DOM 中最高层次的对象,是 DOM 这棵树的根,window 对象使用时可以省略不写,如 window.location.href="要转到的 url",可以直接写为 location.href="要转到的 url"。window 对象提供了定时器方法,可以用这些方法运行定制的定时函数,下面介绍这些定时的方法。

1. window.setTimeout()

用于设置调用一个函数或者其他语句前所等待的时间。

语法格式：

`setTimeout(myFunction,ms)`

浏览器打开窗口后，经过指定时间（ms 毫秒），开始运行 myFunction。浏览器执行下面的代码，3 秒后会弹出消息框。

```
<script>
    setTimeout(msg, 3000);
    function msg()
    {
        alert("时间到");
    }
</script>
```

2. window.clearTimeout()

清除 window.setTimeout() 设置的需要等待的时间。用法如下：

```
var myTimeout = setTimeout(myfunction, 3000);
clearTimeout(myTimeout);
```

下面示例一直在文本框中填充"!"，直到单击 stop 命令按钮。

```
<body>
<input type = "text" id = "msg" />
<input type = "button" value = "stop" id = "bb" onclick = "bb()"/>
<script>
    var iTime;
    function aa()
    {
        msg.value += "!";
        iTime = setTimeout("aa()",1000);
    }
    function bb()
    {
        clearTimeout(iTime);
    }
    aa()
</script>
</body>
```

3. window.setInterval()

用于设置每经过指定时间，就重复调用某函数或其他语句。

语法格式：

`setInterval(repeatFunction,ms)`

指定每经过单位毫秒（ms）就调用一次 repeatFunction 函数。以下代码每经过 1 秒，显示一次时间：

```
<body>
    <div id="showTime">这时显示时期和时间</div>
    <script>
        setInterval(clock, 1000)
        function clock() {
            var t = new Date();
            document.getElementById("showTime").innerHTML = t;
        }
    </script>
</body>
```

4. window.clearInterval()

用 setInterval()设置的定时器会一直运转下去,除非使用 clearInterval()停止计时器。使用方法如下:

```
var myInterval = setInterval(repeatFunction,ms);
clearInterval(myInterval);
```

3.7 XML

XML(eXtensible Markup Language,可扩展标记语言)被设计用来传输和存储数据。XML 是可扩展标记语言,其设计宗旨是传输数据,而不是像 HTML 用于显示数据,XML 标签没有被预定义,用户根据需要自行定义。HTML 不区分大小写,但 XML 区分大小写。XML 是对 HTML 的补充,而不是要取代 HTML。下面是一个 book.xml 的文档。

```xml
<?xml version="1.0" encoding="utf-8"?>
<books>
    <book>
        <classification>computer</classification>
        <bname>Head First jQuery</bname>
        <author>Ryan Benedetti</author>
        <price>78.0</price>
    </book>
    <book>
        <classification>computer</classification>
        <bname>精通 C#</bname>
        <author>Andrew Troelsen</author>
        <price>159.0</price>
    </book>
    <book>
        <classification>literature</classification>
        <bname>红楼梦</bname>
        <author>曹雪芹</author>
        <price>60.0</price>
    </book>
    <book>
        <classification>literature</classification>
        <bname>三国演义</bname>
```

```
        <author>罗贯中</author>
        <price>45.0</price>
    </book>
</books>
```

XML 的语法规则主要如下。

（1）<? xml version="1.0" encoding="utf-8"? >是 XML 的声明文件。如果 XML 文档中有中文，encoding 需要设置为"utf-8"；否则会出现乱码。

（2）XML 必须包含根标记，它是所有其他标记的父标记，book.xml 中 books 就是根标记。

（3）所有 XML 标记必须成对出现，若有开始标记，则必须有结束标记，如<bname>必须配对</bname>使用。

（4）XML 区分大小写。

（5）所有标记嵌套时，标记的开始标记与结束标记间不能出现交叉。例如：

```
<b><i>This text is bold and italic</i></b>
```

不能写为：

```
<b><i>This text is bold and italic</b></i>
```

（6）XML 的属性值必须加引号。

（7）在 XML 中，空格会被保留。

XML 文件是文本文件，可以在记事本中直接编写。编写完成将文件扩展名保存为.xml 即可。各种服务器程序如 PHP、ASP.NET、JSP 等都可以生成 XML 文件。ASP.NET 中，利用 ADO.NET 的 DataSet，可以读写 XML。

3.8　JavaScript Object Note

JavaScript Object Note(JSON)是一种轻量极的数据交换格式。JSON 数据格式容易读写，计算机也容易解析和生成，非常适合于建立数据结构和传输数据。JSON 传输数据要比 XML 更加高效。对 Ajax(Asynchronous JavaScript and XML，异步 JavaScript 和 XML)而言，JSON 比 XML 更快且更容易使用。

图 3-10 是使用 Ajax 从服务器请求数据的过程。客户端向服务器请求数据时，如查询，将查询的内容先串行化处理，以便 Ajax 将表单的数据作为一个数据包发送给服务器。数据包一般是 JSON 字符串，传递到服务器后，服务器解析此 JSON 字符串，生成 JSON 对象，解析出相关信息，如查询的关键字，之后向服务器发出查询请求，服务器将查询结果生成 JSON 字符串，然后传给浏览器，浏览器由 JavaScript 解析这个 JSON 字符串，再生成 JSON 对象，显示给用户。

3.8.1　JSON 数据格式

假设有以下关于城市的信息：

```
Name:Beijing
```

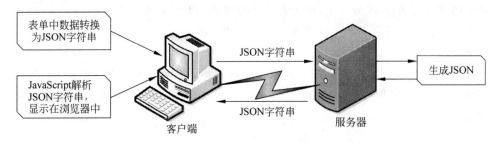

图 3-10　JSON 数据传输过程

Street:Xueyuan Road
Postcode:100083

将以上信息用 JSON 可表示为：

{ Name: "Beijing",Street: "Xueyuan Road", Postcode: "100083" }

JSON 的语法规则如下。

(1) 数据在名称/值对中。如 Name:"Beijing"名称/值,名称和值间用冒号分开。JSON 中的值可以是数字、字符串、逻辑值 true 或 false、放在[]中的数组、放在{}中的对象、null。

(2) 数据由逗号分隔。

(3) 用花括号保存对象。

(4) 用方括号保存数组。

JSON 的结构有两种,即对象和数组。

对象结构以"{"开始,以"}"结束,中间部分由多个以","分隔的"key(关键字)/value(值)"对构成,关键字和值之间以":"分隔,语法结构如下：

```
{
    key1:value1,
    key2:value2,
    ...
}
```

其中关键字是字符串,而值可以是字符串、数值、true、false、null、对象或数组。

数组结构以"["开始,以"]"结束,中间由多个以","分隔的值列表组成,语法结构如下：

```
[
    {
        key1:value1,
        key2:value2
    },
    {
        key3:value3,
        key4:value4
    }
]
```

有两个概念有必要区分一下,即 JSON 字符串和 JSON 对象。

JSON 字符串:符合 JSON 格式要求的字符串,如 var jsonStr= "{carBrand:'Ford', color:'red'}";。

JSON 对象:符合 JSON 格式的对象,如 var jsonObj = {carBrand:'Ford',color:'red'};。

如果要用 JSON 保存两个城市的 Name、Street、Postcode 的信息,则 JSON 数据格式为:

例 3-5 以下面 JSON 数据为例,在 JavaScript 中建立和使用 JSON 对象。

```
<body>
    <div>
        城市:<span id = "cityname"></span>
        街道:<span id = "street"></span>
        邮编:<span id = "postcode"></span>
    </div>
<script>
    var json = {
        Cities:{
            City: [
                    {   Name: "Beijing",
                        Street: "Xueyuan Road",
                        Postcode: "100083"
                    },
                    {   Name: "Shanghai",
                        Street: "Baoshan Road",
                        Postcode: "200071"
                    }
                ]
            }
        };
document.getElementById("cityname").innerHTML = json.Cities.City[1].Name;
document.getElementById("street").innerHTML = json.Cities.City[1].Street;
document.getElementById("postcode").innerHTML = json.Cities.City[1].Postcode;
</script>
</body>
```

程序运行后的输出结果为:

城市:Shanghai 街道:Baoshan Road 邮编:200071

说明:

① 要访问 JSON 对象中的信息,方法与访问其他对象的相同,使用如 json 对象.Cities 的格式。

② JSON 对象中的数组与 JavaScript 中的其他数组一样,具有 length 等相同的属性,例 3-5 中可以使用 json.Cities.City.length 获得共有几个城市。

3.8.2 JSON 文本串转换为 JavaScript 对象

JSON 最常见的用法之一是从 Web 服务器上读取数据,在接收服务器数据时,一般是字符串,需要将其转换为 JavaScript 的对象,然后在网页中使用这些数据。

假设从服务器传递过来的字符串为 '{ Name:"Beijing", Street:"Xueyuan Road", Postcode: "100083" }',转换的方法有以下两种。

(1) 使用 JavaScript 的函数 eval()。

```
<body>
    <div>
        城市:<span id="cityname"></span>
        街道:<span id="street"></span>
        邮编:<span id="postcode"></span>
    </div>
    <script>
        var txt = '{ Name: "Beijing",Street: "Xueyuan Road", Postcode: "100083" }';
        var json = eval("(" + txt + ")");
        document.getElementById("cityname").innerHTML = json.Name;
        document.getElementById("street").innerHTML = json.Street;
        document.getElementById("postcode").innerHTML = json.Postcode;
    </script>
</body>
```

该方法转换时必须将要转换的 JSON 字符串包围在括号中。由于 eval() 函数可编译并执行任何 JavaScript 代码,使用时有潜在的安全问题。

(2) JSON.parse()。

```
<script>
var txt = '{ "Name": "Beijing","Street": "Xueyuan Road","Postcode": "100083" }';
var json = JSON.parse(txt);
        document.getElementById("cityname").innerHTML = json.Name;
        document.getElementById("street").innerHTML = json.Street;
        document.getElementById("postcode").innerHTML = json.Postcode;
</script>
```

说明: 使用 JSON.parse() 时需要注意:①JSON 的字符串中属性名都必须加双引号,如 Street 必须是"Street";②字符串必须是单引号套双引号,如果写作:var txt = "{ 'Name': 'Beijing','Street': 'Xueyuan Road', 'Postcode': 100083 }"; var json = JSON.parse(txt);,则不能转换生成 JSON 对象。

3.8.3 将 JavaScript 对象转换为 JSON 字符串

将一个 JavaScript 对象转换为一个 JSON 字符串,使用 stringfy 方法,语法格式如下:

JSON.stringify(value [, replacer] [, space])

其中,value 是必选字段,是要转换的对象,如数组、类等;replacer 可选,它又分为两种方式,一种是数组,另一种是方法;space 为分隔符。

示例代码如下:

```
<script>
    var student = new Object();
    student.name = "张平";
    student.age = 25;
    student.location = "中国";
    var json = JSON.stringify(student);
    alert(json);
</script>
```

3.8.4 ASP.NET 中浏览器和服务器通过 JSON 的数据交换过程

下面是一个简单浏览器与服务器通过 JSON 完成数据交换的过程,本示例需要 jQuery 库的支持。程序运行结果见图 3-11。

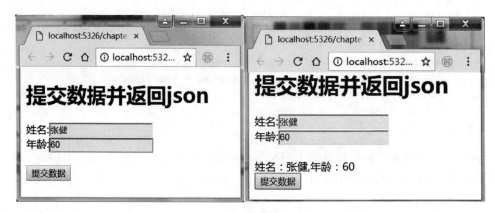

图 3-11 提交数据并返回 JSON

❶ 建立一个 3_8.html 文档,内容如下:

```
1  <!DOCTYPE html>
2  <html xmlns = "http://www.w3.org/1999/xhtml">
3  <head>
4  <meta http-equiv = "Content-Type" content = "text/html; charset = utf-8"/>
5      <title></title>
6  </head>
7  <body>
8      <script type = "text/javascript" src = "../Scripts/jquery.min.js"></script>
9      <h1>提交数据并返回 json</h1>
10     <div id = "main">
```

```
11        <form id="ff">
12            <span>姓名:<input type="text" name="sname"/></span><br/>
13            <span>年龄:<input type="text" name="age"/></span>
14        </form>
15        <div id="content">
16        </div>
17        <button onclick="btnShow()">提交数据</button>
18      </div>
19      <script>
20      function btnShow() {
21          var sendData = $("#ff").serialize();
22          $.get("3_8.ashx", sendData, function (data, status) {
23          var jsonObj = eval('(' + data + ')');
24          $("#content").html("姓名:" + jsonObj.sname + ",年龄:" + jsonObj.age);
25          });
26      }
27      </script>
28    </body>
29 </html>
```

说明:

① 第 8 行将 jQuery 库引入网页中,引用时要注意网页文件与 query.min.js 文件所在的相对位置。

② 第 11 行<form>中没有设置 action 属性,表单以 Ajax 方式提交。

③ 第 17 行在命令按钮增加 onclick 事件,当单击该按钮时执行函数 btnShow(),采用的是 JavaScript 增加事件的方法,jQuery 也有自己增加事件的方法。

④ Ajax 技术允许更新 Web 页面的一部分而无须重新加载整个页面。使用 Ajax 向服务器发送数据前,将发送的数据串行化为一个对象,这样 Ajax 调用就可以将其作为一个数据包发送。jQuery 提供了两个表单辅助方法来完成数据的串行化,即 serialize 和 serializeArray。serialize 将表单输入连接起来组成一个键/值对组成的串,各个键/值对之间用 & 号分隔。serializeArray 将表单输入组成一个键/值对关联数组,是一个对象。与 serialize 相比,serializeArray 数据结构更加清晰;输入姓名为"赵三"、年龄 50 后,sendData 存储的值为 sname=%E8%B5%B5%E4%B8%89&age=25,URL 编码时中文显示为乱码。

⑤ 第 22 行使用 jQuery 的 $.get 方法,发送一个 HTTP GET 请求到服务器并返回结果,参数意义如下:"3_8.ashx"表示要请求的处理程序,sendData 是请求处理程序时将这些数据发给处理的程序,function (data, status)是请求成功后运行的函数,data 是返回的值,status 是请求的状态,取值是"success"、"notmodified"、"error"、"timeout"、"parsererror"。也可以使用 $.post()方法发送数据到服务器,如果请求的是 JSON 格式的数据,还可以使用 $.getJSON()方法。这些方法在 jQuery 中有详细的介绍。

请求处理的程序,这里采用的是.ashx(一般处理程序),而没有采用.aspx(Web 窗体程序)。.ashx 程序适合产生供浏览器处理的、不需要回发处理的数据格式,如用于生成动态图片、动态文本等内容。当然这里采用.aspx 也可以,只是有些浪费资源。

⑥ 3_8.ashx 接收到网页请求后,生成一个 JSON 的字符串,第 23 行将该字符串转变为 JavaScript 的对象。

⑦ 第 24 行 jQuery 的 html() 方法,也可以改为 JavaScript 的语法:

document.getElementById("content").innerHTML = "姓名: " + jsonObj.sname + ",年龄: " + jsonObj.age;

❷ 建立一个 3_8.ashx 文档。

```
using System;
using System.Collections.Generic;
using System.Linq;
using System.Web;
using System.Text;
namespace aspExample.chapter3
{
    public class _3_8 : IHttpHandler
    {
        public void ProcessRequest(HttpContext context)
        {
            context.Response.ContentType = "text/plain";
            string sname = context.Request["sname"];
            string age = context.Request["age"];
            StringBuilder json = new StringBuilder();
            json.Append("{\"sname\":\"");
            json.Append(sname);
            json.Append("\",\"age\":");
            json.Append(age);
            json.Append("}");
            context.Response.Write(json);
        }
        public bool IsReusable
        {
            get
            {
                return false;
            }
        }
    }
}
```

说明:

① Request 是 ASP.NET 提供的内部对象,是 System.Web.HTTPRequest 类的实例,这个对象代表了引起页面被加载的 HTTP 请求的值和属性,它包含所有的 URL 参数以及其他所有的由客户端发送的信息。服务器接收客户端发送的信息,语法格式为 Request["xx"],其中 xx 是 URL 后面的变量名或者是表单中控件的 name 属性的值。在一般处理程序.ashx 中,使用方法是 context.Request["xx"]。

② 拼接 JSON 字符串时,使用 StringGuilder 类时,需要引入命名空间 System.Text;在 json.Append("{\"sname\":\"")中为了拼接双引号,需要使用\转义字符。很显然,如果要

输出的 JSON 字符串比较复杂时,手工拼写字符串很麻烦而且容易出现错误。为此 C♯中提供了序列化和反序列化的方法:将 JSON 对象序列化为 JSON 字符串;将 JSON 字符串反序列化为 JSON 对象。这些内容将在第 5 章有详细论述。

本示例只是简单地演示了客户端与服务器 JSON 数据交换的过程。实际编程中服务器需要调用数据库,将查询的结果生成 JSON 字符串,传递到客户端,客户端利用 JavaScript、jQuery 解析此 JSON 字符串,然后在浏览器中显示出查询结果。

习题与思考

(1) 将例 3-2 中内部样式表改为外部样式表,存为 myStyle.css,然后将该外部样式表引用到例 3-2 中。

(2) 用 DIV+CSS 制作一个页面,浏览器中显示效果如图 3-12 所示。

图 3-12　浏览器显示效果

(3) 程序运行后,单击"显示全部记录"按钮,显示如图 3-13 所示,请填空。

图 3-13　显示全部记录

阅读以下 JavaScript 程序，请填空。

```html
<html>
<body>
    <div id="display">
    </div>
    <button onclick="execute()">显示全部记录</button>
    <script>
        var people = [
            new person('1',"张拴柱","男","2010-10-05"),
            new person('2',"赵铁锤","男","2015-04-30"),
            new person('3',"李梅梅","女","2012-12-20"),
            new person('4',"张欣芳","女","2000-08-20")
        ]
        function person(no, name, gender, birthday) {
            _____;
            _____;
            _____;
            _____;
        }
        function execute()
        {
            for (var i = 0; _____; i++)
            {
                document.getElementById("display")._____ += "编号: " + people[i].no + "<br>"
                    + "姓名: " + _____ + "<br>";
            }
        }
    </script>
</body>
</html>
```

（4）按照"性别"显示"学生名单"，浏览器显示结果如图 3-14 所示。

图 3-14　浏览器显示结果

单击"按性别显示名单"命令按钮，显示如图 3-15 所示。
阅读以下 HTML 和 JavaScript 代码，请填空。

```
<!DOCTYPE html>
<html xmlns="http://www.w3.org/1999/xhtml">
<head>
```

图 3-15　显示结果

```
<meta http-equiv="Content-Type" content="text/html; charset=utf-8"/>
    <title></title>
</head>
<body>
<div id="female">
    女生名单:
</div>
<div id="male">
    男生名单:
</div>
<button onclick="display()">按性别显示名单</button>
<script>
    function display() {
        var json = {
            "students": {
                "student": [
                    { "sno": 30012035, "sname": "楚云飞", "ID": "230304200611201537" },
                    { "sno": 30012036, "sname": "张雨", "ID": "230304200210052346" },
                    { "sno": 40012030, "sname": "赵三", "ID": "230304200010302378" },
                    { "sno": 41321059, "sname": "李孝诚", "ID": "110106199802030013" },
                    { "sno": 41340136, "sname": "马小玉", "ID": "130304199705113028" },
                    { "sno": 41355062, "sname": "王长林", "ID": "110106199512035012" },
                    { "sno": 41361045, "sname": "李将寿", "ID": "14510619901203503X" },
                    { "sno": 41361258, "sname": "刘登山", "ID": "130030520001234235" },
                    { "sno": 41361260, "sname": "张文丽", "ID": "140305200106202348" },
                    { "sno": 41401007, "sname": "鲁宇星", "ID": "110108200012035012" }
                ]
            }
        };
        document.getElementById("female").innerHTML = "<font color=red>女生名单:</font><br>"
        document.getElementById("male").innerHTML = "<font color=red>男生名单:</font><br>"
```

```
            for(var i = 0;i<_____;i++)
            {
                if (_____)
                {
                    _____;
                }
                else
                {
                    _____;
                }
            }
        }
    </script>
    </body>
</html>
```

提示:"性别"可利用 JavaScript 中的 substr()函数,通过截取 ID 字符串加以判别。

(5)下面是关于天气 API 接口返回的天气预报的 JSON,请用 JavaScript 解析并在网页中显示:

{"weatherinfo":{"city":"北京","cityid":"101010100","temp1":"24℃","temp2":"11℃","weather":"雷阵雨转多云","img1":"d4.gif","img2":"n1.gif","ptime":"11:00"}}

ASP.NET内置对象

（1）Page 对象的属性和事件。
（2）Response 对象的主要属性方法。
（3）Request 对象的主要属性方法。
（4）Session 对象的主要属性方法。
（5）Application 对象的主要属性方法。
（6）Server 对象的主要属性方法。

Web 程序运行时，ASP.NET 将一些常用的类的实例已经自动创建好了，用户可以直接使用这些对象，而无须再通过类的实例化创建这些对象。已经创建好的对象就是 ASP.NET 的内置对象。表 4-1 列出了这些常用的内置对象。

表 4-1 ASP.NET 常用的内置对象

对象名	描 述	ASP.NET 类
Response	提供对当前页面输出流的访问	HttpResponse
Request	提供对当前页请求的访问，包括请求标题、Cookies、客户端证书、查询字符串等。可以使用该类读取浏览器发出的内容	HttpRequest
Session	为当前用户的会话提供信息，在会话范围内，还提供用于存储信息缓存的访问以及控制如何管理会话的方法	HttpSessionState
Application	在应用程序范围内，提供对所有会话的方法和事件的访问，还提供对用于存储信息缓存的访问	HttpApplicationState
Server	公开可以用于在页之间传输控件的实用工具方法，获取有关最新错误的信息，对 HTML 文本进行编码和解码等	HttpServerUtility
Context	提供对整个当前上下文（包括请求对象）的访问，可以使用此类共享页之间的信息	HttpContext
Trace	提供在 HTTP 页输入中显示系统和自定义跟踪诊断所有消息的方法	TraceContext
Page	ASP.NET 网页中包含的所有服务器控件的一个最外围的命名容器	

4.1 Page 对象

所有的 Web 页面都继承自 System.Web.UI.Page 类,Page 类与扩展名为.aspx 的文件相关联,这些文件在运行时被编译为 Page 对象,并缓存在服务器的内存中。ASP.NET 页面中所有的控件总是属于 Page 对象,Page 对象常用的属性有 Title、IsPostBack、IsValid 等。

1. IsPostBack

该属性是 Page 中最重要的属性,使用该属性可判断是首次加载访问该页面,还是为响应客户端回发而加载。首次加载页面时,该页面的 IsPostBack 属性是 false,响应客户端回发时,该属性为 true。

例 4-1 新建立一个 Web 窗体,窗体上放置标签控件 Label1 和 Label2、文本框 TextBox1、命令按钮 Button1。运行窗体后,命令按钮显示为图 4-1(a),在文本框中输入内容后,单击"提交"按钮后,变为图 4-1(b)。

图 4-1 IsPostBack 示例

```
protected void Page_Load(object sender, EventArgs e)
{
    if (Page.IsPostBack == true)
    {
        Label2.Text = "你填写的姓名是:" + TextBox1.Text;
    }
    else
    {
        Label2.Text = "请填写姓名";
        TextBox1.Focus();
    }
}
```

说明:首次访问该页面时,Page.IsPostBack 为 false,故 Label2 显示"请填写姓名";单击命令按钮后,网页处于回传状态,Page.IsPostBack 为 true。

2. isValid 属性

该属性用于获取一个布尔值,指示网页上的验证控件是否验证成功,如果网页上的验证控件全部验证成功,则该值为 true;否则为 false。下面结合 ASP.NET 中的验证控件,说明该属性的使用。

例 4-2 Web 窗体上通过服务器端验证"用户名"是否重复,说明验证控件及 isValid 属性的使用。

❶ 在 SQL Server 的 students 数据库中,建立一个表 users,表结构及表中记录见图 4-2 和图 4-3。

❷ 新建立一 Web 窗体 4_2.aspx，窗体布局页面见图 4-4，页面中属性设置见表 4-2。

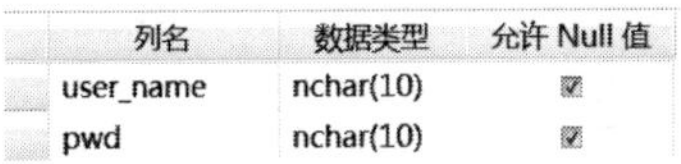

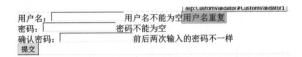

列名	数据类型	允许 Null 值
user_name	nchar(10)	☑
pwd	nchar(10)	☑

图 4-2　users 表结构

	user_name	pwd
1	admin	admin

图 4-3　users 中的记录

图 4-4　页面 4_2.aspx 的布局

表 4-2　页面 4_2.aspx 各控件属性的设置

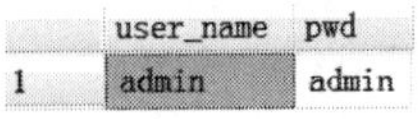

项　　目	控件类型	控件 ID	其他属性的设置
用户名	TextBox	txtUserName	TextMode：Single
密码	TextBox	txtPwd	TextMode：Password
确认密码	TextBox	txtPwdAgain	TextMode：Password
用户名验证控件 1	RequiredFieldValidator	RequiredFieldValidator1	ControlToValidate：txtUserName ErrorMessage：用户名不能为空
用户名验证控件 2	CustomValidator	CustomValidator1	ControlToValidate：txtUserName ErrorMessage：用户名重复
密码验证控件	RequiredFieldValidator	RequiredFieldValidator2	ControlToValidate：txtPwd ErrorMessage：密码不能为空
确认密码控件	CompareValidator	CompareValidator1	ControlToCompare：txtPwd ControlToValidate：txtPwdAgain ErrorMessage：前后两次输入的密码不一样
提交	Button	txtSubmit	Text：提交

❸ 在图 4-4 中双击 CustomValidator1，进入该自定义控件的方法代码编写窗口，编写服务器端验证代码如下：

```
protected void CustomValidator1_ServerValidate(object source, ServerValidateEventArgs args)
{
 string userName = args.Value;
 string sqlStr = "select count( * ) from users where user_name = '" + userName + "'";
 SqlConnection cnn = new SqlConnection();
 cnn.ConnectionString = @"server = .;database = students;uid = sa;pwd = 3500";
 cnn.Open();
 SqlCommand cmd = new SqlCommand(sqlStr,cnn);
 int num = (int)cmd.ExecuteScalar();
```

```
        cnn.Close();
        if (num == 0)
        {
            args.IsValid = true;              //验证通过
        }
        else
        {
            args.IsValid = false;
        }
    }
```

❹ 图 4-2 中双击"提交"命令按钮,进入该按钮单击事件代码编写窗口,编写以下代码:

```
protected void btnSubmit_Click(object sender, EventArgs e)
{
    if (Page.IsValid == true)
    {
        Response.Write("验证通过");
    }
    else
        Response.Write("验证没有通过");
}
```

本示例运行前,需要引入命名空间 using System.Data.SqlClient,程序运行后,如果用户名中输入的是 admin,会提示"用户名重复"。

3. Page 对象的主要事件

Page 对象具有 Load、Init、PreLoad、PreInit 等 8 个主要事件,最常用的是 Page_Load 事件。该事件的两个参数是由 ASP.NET 定义的,第 1 个参数定义了产生事件的对象,第 2 个参数是传给事件的详细信息。每次触发服务器控件时,页面都会执行一次 page_Load 事件,说明页面被加载了一次,这个技术称为回传技术。在 ASP.NET 中,当客户端触发了一个事件,它不是在客户端浏览器上对事件进行处理,而是将该事件的信息传送回服务器进行处理。服务器在接收到这些信息后,会重新加载 Page 对象,然后处理该事件,故 Page_Load 被再次触发。

由于 Page_Load 在每次页面加载时运行,因此其中的代码即使在回传的情况下也会被执行,此情况下,Page 的 IsPostBack 属性就可以解决此问题,用法见例 4-1。

4.2 Response 对象

Response 对象是 System.Web.HttpResponse 类的实例,Response 对象封装了 Web 服务器对客户端请求的响应,它用来操作 HTTP 相应的信息,用于将结果返回给请求者,虽然 ASP.NET 中控件的输出不需要去写 HTML 代码,但很多时候需要手动控制输出流,如文件的下载、重定向、脚本输出等。

4.2.1 Response 对象的属性和方法

Response 对象常用的属性见表 4-3,方法见表 4-4。

表 4-3 Response 对象的属性

属性	描述
Buffer	获取或者设置一个值,该值指示是否缓冲输出,并在完成处理整个响应后将其发送
Expires	获取或者设置在浏览器上缓存的页过期之前的分钟数,如果用户在页过期前返回同一页,则显示缓存的版本
Cookies	设置 Cookies 集合

表 4-4 Response 对象的方法

方法	描述
Write	将指定的字符串或者表达式写到 HTTP 输出中
End	停止页面的执行并得到响应的结果
Clear	在不将缓存中的内容输入的情况下,清空当前页的缓存。需要设置 Response.Buffer=true 后再使用该方法
Flush	将缓存中的内容立即输出。需要设置 Response.Buffer=true 后再使用该方法
Redirect	使浏览器重新定向到指定的 URL,经常用于访问页面权限控制上

Response.Write()输出指定的字符串或者表达式时,输出的字符串中如果包含 HTML 标记,浏览器会将其执行后输出。例如:

Response.Write("< p style = 'color:red'>北京科技大学</p>");

将输出红色字体的"北京科技大学"。

Response.Write()中常采用下列方式,在浏览器中输出一个 alert()对话框:

Response.Write("< script >alert('对话框的内容')</script >");

4.2.2 Response 对象应用示例

例 4-3 有 login.aspx 和 secretContent.aspx 两个页面,只有在 login.aspx 页面的文本框 txtUserName 中输入 admin 后,单击"确认"按钮,页面转到 secretContent.aspx 页面,并输出"你是 admin,可以浏览本页面";如果在浏览器中直接输入要访问 secretContent.aspx 的 URL,浏览器会自动转到 login.aspx 页面。

❶ 新建一个 Web 窗体 login.aspx,页面布局如图 4-5 所示。

❷ 设置文本框的 ID 为 txtUserName,双击命令按钮 Button1,进入代码编辑窗口,写入以下代码:

图 4-5 login.aspx 页面

```
protected void Button1_Click(object sender, EventArgs e)
{
    if (txtUserName.Text == "admin")
    {
        Response.Cookies["username"].Value = "admin";
        Response.Redirect("secretContent.aspx");
    }
}
```

说明:Response.Cookies["username"].Value = "admin";将字符 admin 保存在键值

为 username 的 Cookies 集合中,以便在其他页面中使用该值。Cookies 的用法见后面的 Cookie 对象。

❸ 新建一个 Web 窗体 ecretContent.aspx,双击该页面,进入代码编辑窗口,编写以下代码:

```
protected void Page_Load(object sender, EventArgs e)
{
    if (Request.Cookies["username"] == null)
        Response.Redirect("login.aspx");
    else
    {
        if (Request.Cookies["username"].Value == "admin")
        {
            Response.Write("你是 admin,可以浏览本页面");
        }
        else
            Response.Redirect("login.aspx");
    }
}
```

说明:

① if(Request.Cookies["username"] == null)判断 Cookies 集合中是否存在键值为 username 的 Cookie;Request 对象的应用见后面。

② if(Request.Cookies["username"].Value == "admin"),如果 Cookies["username"]的值是 admin,利用 Response.Write 输出"你是 admin,可以浏览本页面";否则利用 Response.Redirect("login.aspx")将浏览器重新定向到"login.aspx"页面。

❹ 运行 secretContent.aspx,浏览器会转到 login.aspx 页面,只有在文本框输入 admin,secretContent.aspx 页面才会输出"你是 admin,可以浏览本页面"。

Redirect()方法需要在客户端和服务器端往返一次。本质上它给浏览器发送一条消息,告诉浏览器跳转到一个新页面。该方法有一个以 true 或 false 为第 2 个参数的重载,该参数表示跳转后是否让页面继续运行,页面的其他事件也将继续运行。如果第 2 个参数设置为 true,ASP.NET 会立即停止处理页面,减少 Web 服务器的负荷。

如果只是在同一个 Web 应用程序里向用户发送另一个 Web 表单,可以使用更快的 Server.Transfer()方法。使用 Server.Transfer()的缺陷是,由于重新定向只是发生在服务器端,客户端浏览器保持了原有的 URL,相应地,浏览器没有办法知道目前显示的是一个不同的页面,如果客户端刷新页面或者要收藏页面,就会出现问题。另外,Server.Transfer()不能将执行传送到非 ASP.NET 的页面、其他 Web 服务器或者其他 Web 应用程序的网页中。

4.3 Request 对象

Request 对象是 System.Web.HttpRequest 类的一个实例。当客户端从网站请求 Web 页面时,Web 服务器接收一个客户端的 HTTP 请求,客户端的请求信息会包装在 Request 对象中。这些请求信息包括报头(Header)、客户端的主机信息、客户端浏览器信息、请求方法(如 POST、GET)和提交的窗体信息等。

Request 对象的主要属性和方法见表 4-5 和表 4-6。

表 4-5 Request 对象的主要属性

属 性	描 述
QueryString	获取 HTTP 查询字符变量集合
Form	获取表单值的集合
UserHostAddress	获取远程客户端的 IP 主机地址
Browser	获取正在请求的客户端浏览器功能的信息
ServerVariables	获取正在请求的客户端信息的集合,如客户端的 IP 等
Cookies	获取 Cookies 集合

表 4-6 Request 对象的主要方法

方 法	说 明
BinaryRead	执行对当前输入流进行指定字节数的二进制读取
MapPath	将当前请求的 URL 中的虚拟路径映射到服务器上的物理路径

4.3.1 Form 集合

用于获取表单提交的数据。

例 4-4 设计 4_4a.aspx 和 4_4b.aspx 两个页面,4_4a.aspx 用于输入数据,4_4b.aspx 获取 4_4a.aspx 中输入的数据。4_4a.aspx 界面设计如图 4-6 所示,图中两个文本框的 ID 分别为 txtNo、txtUserName,命令按钮的 ID 为 Button1。

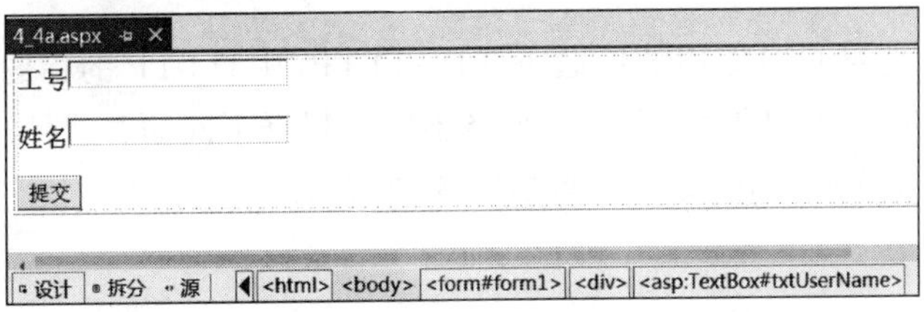

图 4-6 输入界面

❶ 新建 Web 窗体 4_4a.aspx,设计图 4-6 所示界面,设置两个文本框的 ID 属性为 txtNo、txtUserName。

❷ 图 4-6 中单击"源"视图,将<form>标记更改为:

<form id="form1" runat="server" method="post" action="4_4b.aspx">

在该网页中不编写 Button1_Click 事件过程,Button1 仅仅起到提交网页的作用。

❸ 新建 Web 窗体 4_4b.aspx,该页面中仅包含一个标签 Label1,页面在"设计"下,双击该页面,编写以下事件过程:

```
protected void Page_Load(object sender, EventArgs e)
{
    string userNo = Request.Form["txtNo"];
```

```
    string userName = Request.Form["txtUserName"];
    Label1.Text = "你输入的工号是：" + userNo + ",姓名是：" + userName;
}
```

❹ 运行 4_4a.aspx，输入"工号"和"姓名"后，单击"提交"按钮。

说明：Request.Form["txtNo"] 获取 4_4a.aspx 中的 ID 是 txtNo 的值。Request.Form["txtNo"] 也可以简写为 Request["txtNo"]，Request 对象获取集合数据的先后顺序是 QueryString、Form、Cookies、ServerVariables，在获取数据时，加上这些集合的名称，可优化程序的执行效率。

4.3.2 QueryString 集合

在程序中经常使用 QueryString 来获取从一个页面传递过来的字符串参数。例如，在一页面 a 中建立一个链接，指向页面 b，并传递给页面 b 两个变量：

```
<a href="b.aspx?userNo=5&userName=张三">查看</a>
```

在页面 b.aspx 中使用 QueryString 接收从页面 a.aspx 传递过来的两个变量：

```
protected void Page_Load(object sender, EventArgs e)
{
    string userNo = Request.QueryString["userNo"];
    string userName = Request.QueryString["userName"];
}
```

同样地，上面的 Request.QueryString["userNo"] 也可以写为 Request["userNo"]。

4.3.3 ServerVariables 集合

在浏览器中浏览网页时使用的传送协议是 HTTP，在 HTTP 的标题文件中会记录一些客户端的信息，如客户的 IP 地址等，有时服务器需要根据不同的客户信息做出不同的反应，这时就需要用 ServerVariables 集合获取所需要的信息。

语法格式：

```
Request.ServerVariables["服务器环境变量"];
```

部分服务器环境变量见表 4-7。

表 4-7 服务器环境变量

环境变量名	描述
Server_Name	获取服务器的主机名、DNS 地址或 IP 地址
HTTP_Referer	获取访问来源页面
OS	获取操作系统
Appl_Physical_Path	获取 IIS 物理路径
Script_Name	获取虚拟路径
Server_Port	服务器处理请求的端口
Path_Info	客户端提供的路径信息
Script_Name	执行脚本的名称

续表

环境变量名	描述
Remote_Addr	发出请求的远程主机的 IP 地址
Remote_Host	发出请求的远程主机名称
Request_Method	提出请求的方法,如 GET、HEAD、POST 等
Server_Protocol	服务器使用的协议名称和版本
Server_Name	服务器的主机名、DNS 地址或 IP 地址
Server_Software	应答请求并运行网关的服务器软件的名称和版本

如果要输出服务器的 IP,语句为:

Response.Write(Request.ServerVariables["SERVER_NAME"])

4.4 Server

Server 对象是 System.Web.HttpServerUtility 类的实例,它提供若干不同的辅助方法和属性,如表 4-8 所示。

表 4-8 Server 的属性和方法

属性/方法	描述
MachineName	该属性表示运行页面的计算机名称。Web 服务器计算机使用这个名称在网络上标识自身
GetlastError()	获取最近发生的异常对象。这个错误必须是在处理当前请求时发生的,并且不能是已经被处理过的
HtmlEncode()	将普通的字符串转化成合法的 HTML 字符串
HtmlDecode()	将 HTML 字符串转化为普通的字符串
UrlEncode()	将普通的字符串转化成合法的 URL 字符串
UrlDecode()	将 URL 字符串转化为普通的字符串
MapPath()	返回对应于 Web 服务器上特定虚拟文件路径的物理文件路径
Transfer()	转到指定的网页。与 Response.Redirect()相似,但更快。它不能把用户转到 Web 服务器上其他站点或者非 ASP.NET 的页面(如 HTML 页面或者 ASP 页面)

4.4.1 Transfer()方法

Transfer()方法是应用程序中把用户重新定向到其他页面速度最快的方法,使用这个方法不需要在客户端和服务器端之间往返。ASP.NET 引擎直接加载新页面并开始处理该页面,故显示在客户端浏览器的 URL 不会发生变化。

例如,在 4_4b.aspx 的 Page_Load 中增加语句:

Server.Transfer("login.aspx");

运行 4_4b.aspx,页面直接跳转到 login.aspx,但浏览器上 URL 的地址并没有改变,见图 4-7。

图 4-7 页面跳转但 URL 保持不变

4.4.2 MapPath()方法

MapPath()方法是 Server 对象的一个非常有用的方法,假如从当前虚拟文件夹下加载一个文本文件 info.txt,除了使用绝对路径外,还可以使用 Server.MapPath("info.txt")将相对路径转换为绝对路径。例如:

```
String physicalPath = Server.MapPath("inf.txt");
StreamReader reader = new StreamReader(physicalPath);
…
Reader.Close();
```

4.4.3 HTML 和 URL 编码

如果要将以下内容输出到浏览器:

<div>使用<p>段落标记</p></div>

如果直接使用:

Response.Write("使用<b>标记加粗字体");

输出的结果是:

使用标记加粗字体

输出的内容中没有出现<b>,浏览器将<b>解释成把它后面的文字加粗的指令,要避免此种情况的发生,可以使用 Server.MapPath()方法:

Response.Write(Server.HtmlEncode("使用<b>标记加粗字体"));

如果要显示数据库中的记录,在记录的内容可能存在 HTML 的情况下,HtmlEncode()方法就显得非常重要了。类似地,UrlEncode()方法可以通过转义空格及其他特殊字符把文本转换为可以在 URL 中使用的格式。

4.5 ASP.NET 状态管理

Web 应用程序框架是基于一种无序状态的 HTTP 协议,每次 Web 请求后,客户端和服务器端断开,同时 ASP.NET 引擎释放页面对象。这种架构保证了 Web 应用程序能够同时响应数千个并发请求而不会导致服务器内存崩溃,其不足是必须通过其他技术存储 Web 请求间的信息,并在需要时获取这些信息。

ASP.NET 提供了很多状态管理的机制,包括 ViewState 对象、Cookie 对象、Session 对象和 Application 对象。

4.5.1 ViewState 对象

与其他状态管理使用的对象相似,ViewState 采用的是字典集合的方式。字典集合中每个项目通过一个唯一的字符串名字进行索引,如下面的代码:

```
ViewState["Counter"] = 1;
```

将值 1 放入 ViewState 集合中,同时给它起名 Counter,如果集合中没有此名称的索引项,它将自动被添加到集合中;如果集合中已经有了名为 Counter 的索引项,原来的内容将会被替换。

ViewState 集合能够处理不同的数据类型,它将所有的数据保存为 Object 类型,故在读取集合中数据时,还需要转换为适当的数据类型。

ViewState 的限制是必须与特定页面绑定,当用户从一个页面浏览到另一个页面时,这些信息就会消失。

下面是从 ViewState 中读取 Counter 值的代码如下:

```
int counter;
if(ViewState["Counter"]!= null)
{
    counter = ViewState["Counter"];
}
```

从 ViewState 集合中读取数据时,如果集合中不存在该键值,会抛出异常,故在读取前需要判断包含该键值的集合是否为空。

例 4-5 使用 ViewState 保存表单中的数值或字符串,程序运行结果如图 4-8 所示,单击"保存"按钮,将表单中文本框的值保存在 ViewState 中,单击"恢复"按钮,将保存在 ViewState 集合中的值显示在文本框中。

图 4-8 使用 ViewState 保存和恢复文本

❶ 新建 Web 窗体 4_5.aspx,页面上放置两个标签、两个文本框和两个命令按钮。将标签的 Text 属性设置为"工号"和"姓名",命令按钮的 ID 设置为 cmdSave 和 cmdRestore。

❷ 完整的程序代码如下:

```
public partial class _4_5: System.Web.UI.Page
{
    protected void cmdSave_Click(object sender, EventArgs e)
    {
        SaveAllText(form1.Controls);
    }
    private void SaveAllText(ControlCollection controls)
    {
        foreach (Control control in controls)
        {
            if (control is TextBox)
            {
                ViewState[control.ID] = ((TextBox)control).Text;
            }
        }
    }
    private void RestoreAllText(ControlCollection controls)
    {
        foreach (Control control in controls)
```

```
            {
                if (control is TextBox)
                {
                    if (ViewState[control.ID] != null)
                    {
                        ((TextBox)control).Text = (string) ViewState[control.ID];
                    }
                }
            }
        }
        protected void cmdRestore_Click(object sender, EventArgs e)
        {
            RestoreAllText(form1.Controls);
        }
    }
```

说明：由于控件的 ID 在页面上是唯一的，故它被用作 ViewState 的键值。

使用 ViewState 保存对象与保存数值或者字符串一样，比较简单，只是在保存对象时 ASP.NET 需要将对象序列化，以便添加到页面隐藏的字段中。对象在默认情况下是不能被序列化的，要使对象序列化，需要在类的声明前加上 Serialization 特性。例如，下面的 Employees 类：

```
[Serializable]
public class Employees
{
    public string EmployeeNo;
    public string EmployeeName;
    public Employees(string employeeNo, string employeeName)
    {
        employeeNo = this.EmployeeNo;
        employeeName = this.EmployeeName;
    }
}
```

由于 Employees 类被标识为可序列化，所以它可以被保存在 ViewState 中：

```
Employees employeeA = new Employees("A950103","乔峰");
ViewState["CurrentEmployee"] = employeeA;
```

从 ViewState 中获取自定义对象后，在使用前需要进行类型的转换：

```
Employees employeeB = (Employees)ViewState["CurrentEmployee"];
```

例 4-6 使用 ViewState 保存对象，将保存结果加以显示，如图 4-9 所示。单击"保存"按钮，将文本框中的信息保存在 Dictionary 对象中；单击"显示"按钮，将 ViewState 中保存的对象用标签显示出来。

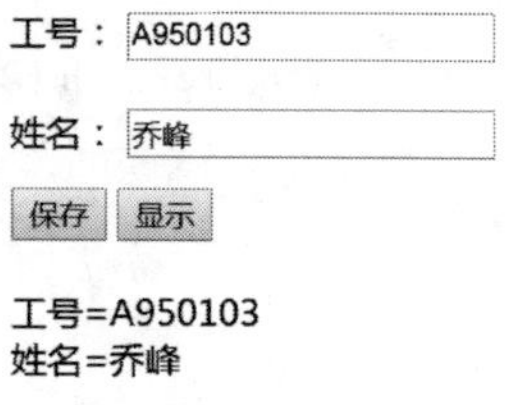

图 4-9 使用 ViewState 保存对象

❶ 新建窗体文件 4_6.aspx，页面上放置两个标签、两个文本框（将其 ID 分别设置为"工

号"和"姓名"),用于输入"工号"和"姓名";两个命令按钮 cmdSave 用于保存 ViewState,cmdDisplay 用于将 ViewState 显示在标签 lblResult 上。

❷ 编写以下各事件中的代码:

```
protected void cmdSave_Click(object sender, EventArgs e)
{
    SaveAllText(form1.Controls);
}

private void SaveAllText(ControlCollection controls)
{
    Dictionary<string, string> myDic = new Dictionary<string, string>();
    foreach (Control control in controls)
    {
        if (control is TextBox)
        {
            myDic.Add(control.ID, ((TextBox)control).Text);
        }
    }
    ViewState["CurrentEmployee"] = myDic;
}
protected void cmdDisplay_Click(object sender, EventArgs e)
{
    if (ViewState["CurrentEmployee"] != null)
    {
        Dictionary<string, string> myDic = (Dictionary<string, string>)ViewState["CurrentEmployee"];
        foreach (KeyValuePair<string, string> item in myDic)
        {
            lblResults.Text += item.Key + "=" + item.Value + "<br/>";
        }
    }
}
```

说明:该示例使用了泛型 Dictionary 类,该类是 system.Collections.Generic 命名空间下提供的一个可序列化的键值集合。只要使用 Dictionary 保存可序列化的对象,并为键使用可序列化的数据类型,就可以把 Dictionary 对象保存到 ViewState 中。关于泛型和 Dictionary 将在第 5 章有详细的介绍。

4.5.2 Cookies

Response 和 Request 对象都有一个 Cookies 属性,它是存放 Cookie 对象的集合,是 HttpCookieCollection 类对象,提供了操作 Cookie 的方法。Cookie 是 HttpCookie 类对象,提供创建和操作各 Cookie 的方法。一个 Cookie 是一段文本信息,能够随着用户请求和网页在 Web 服务器和浏览器之间传递。用户每次访问站点时,Web 应用程序都可以读取 Cookie 包含的信息,从而知道用户上次登录的时间等具体信息。Cookie 可以设置特定的过期时间和日期,也可以是永久的。

大多数浏览器支持最大为 4096B 的 Cookie,从而限制了 Cookie 的大小,故 Cookie 只能用于存储少量的数据,或者存储用户 ID 之类的标识符,根据用户 ID 便可标识用户及从数据库或者其他数据源中读取用户信息。

Cookie 对象和 Application、Session 对象一样,都是为了保存信息,区别在于 Cookie 是保存在客户端,而 Application、Session 是保存在服务器端。

使用 Response 对象的 Cookies 集合属性设置 Cookie 信息,使用 Request 对象的 Cookies 集合属性读取 Cookie 信息。Cookies 集合中有 Count 属性,可返回 Cookie 对象的个数;Add 方法向 Cookies 集合中新增加一个 Cookie 对象;Clear 方法删除 Cookies 集合中所有的 Cookie 对象;Remove 方法删除 Cookies 集合中指定名称的 Cookie 对象。

1. 设置 Cookie

可以通过多种方法将 Cookie 添加到 Cookies 集合中。

方法 1:

```
Response.Cookies["userName"].Value = "admin";
Response.Cookies["userName"].Expires = DateTime.Now.AddDays(1);
```

方法 2:

```
HttpCookie aCookie = new HttpCookie("lastVisit");
aCookie.Value = DateTime.Now.ToString();
aCookie.Expires = DateTime.Now.AddDays(1);
Response.Cookies.Add(aCookie);
```

方法 1 向 Cookies 集合中添加了一个名为 userName 的 Cookies,设置期有效时间是 1 天,该方法 Cookies 集合的值是直接设置的;方法 2 建立了一个 HttpCookie 对象实例 aCookie 并设置其属性,然后通过 Add 方法将其添加到 Cookies 集合中,在实例化 HttpCookie 对象时,必须将该 Cookie 的名称作为构造函数的一部分进行传递。

HttpCookie 类具有以下构造函数:

```
public HttpCookie(string nname)
public HttpCookie(string nname,string value)
```

故在方法 2 中,也可以在建立 HttpCookie 对象实例时,为 Cookie 直接赋值,将前两条语句合并为:

```
HttpCookie aCookie = new HttpCookie("lastVisit", DateTime.Now.ToString());
```

2. 设置多值的 Cookie

一个 Cookie 对象可以有多个值,通过子键区分它们。

例如,当一个名为 userInfo 的 Cookie 对象已经添加到 Response 对象中后,可以通过以下语句设置两个子键的值:

```
Response.Cookies["userInfo"]["userName"] = "admin";;
Response.Cookies["userInfo"]["lastVisit"] = DateTime.Now.AddDays(1).
ToString();
```

或者在创建 Cookie 对象的同时设置多个值：

```
HttpCookie bCookie = new HttpCookie("userInfo");
bCookie.Values["userName"] = "admin";
bCookie.Values["lastVisit"] = DateTime.Now.AddDays(1).ToString();
bCookie.Expires = DateTime.Now.AddMinutes(1);   //1 分钟后 Cookie 过期
```

3. 读取单值的 Cookie

对于单值 Cookie 对象，直接用 Request.Cookies[Cookie 的 Name 值]来读取其 Cookie 值，如设置名为 userName 的单值 Cookie：

```
Response.Cookies["userName"].Value = "admin";
```

读取该 Cookie 值，将其显示在 Label1 中：

```
if (Request.Cookies["userName"] != null)
{
    Label1.Text = Request.Cookies["userName"].ToString();
}
```

在读取 Cookie 前，要先判断 Cookie 是否存在，如果该 Cookie 不存在，直接读取时会出现异常错误。

4. 读取多值的 Cookie

对于多值 Cookie 对象，还需要加上子键名称。如设置名为 userInfo 的 Cookie：

```
Response.Cookies["userInfo"]["userName"] = "admin";
Response.Cookies["userInfo"]["lastVisit"] = DateTime.Now.AddDays(1).ToString();
```

读取该 Cookie 将其结果显示在标签 Label1 和 Label2 上：

```
if (Request.Cookies["userInfo"] != null)
{
    Label1.Text = Request.Cookies["userInfo"]["userName"];
    Label2.Text = Request.Cookies["userInfo"]["lastVisit"];
}
```

5. 设置 Cookie 的有效期

Cookie 的 Expires 属性是 DateTime 类型的，它用来指定 Cookie 的过期日期和时间，即 Cookie 的有效期。浏览器会在适当的时候删除已经过期的 Cookie。如果不指定 Cookie 的有效期，则为会话 Cookie，不会存入用户的硬盘，在浏览器关闭后 Cookie 就会被删除。

6. 修改和删除 Cookie

修改某个 Cookie 实际上是用新的值创建新的 Cookie，并把该 Cookie 发送到浏览器，覆盖客户端上旧的 Cookie。

删除 Cookie 是修改 Cookie 的一种形式。由于 Cookie 存储在客户的计算机中，因此无法将其直接删除，但是可以让浏览器删除。方法是创建一个与要删除的、与 Cookie 同名的、新的 Cookie，并设置该 Cookie 的到期日期早于当前日期的某个日期。当浏览器检查

Cookie的到期日期时,浏览器便会丢弃这个已经过期的Cookie。

例4-7 删除应用程序中所有Cookie。

```
HttpCookie aCookie;
string cookieName;
int cookieCount = Request.Cookies.Count;
for (int i = 0; i < cookieCount; i++)
{
  cookieName = Request.Cookies[i].Name;
  aCookie = new HttpCookie(cookieName);
  aCookie.Expires = DateTime.Now.AddDays(-1);
  Response.Cookies.Add(aCookie);
}
```

4.5.3 Session

Session对象是一个会话对象,是HttpSessionState类的对象,就Web开发而言,一个会话就是客户通过浏览器与服务器间的一次通话。由于HTTP是无状态的,因此无法记录客户一连串的动作,必须有一种机制使服务器能够认得客户,故引入了会话的概念。服务器发送给客户一个会话ID(SessionID),当客户端访问服务器时就带着这个ID,服务器凭借这个唯一的ID来识别客户。当用户请求ASP.NET网页时,系统将自动创建Session对象,在退出应用程序或者关闭服务器时该会话撤销。

Session对象是一个对象的集合,可以看作是存储信息的容器,供本次会话共享。

例4-8 在页面4_8a.aspx中输入用户名和密码,在另一个页面中显示输入的用户名和密码,运行结果见图4-10。

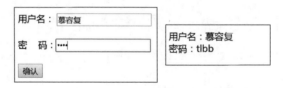

图4-10 使用Session记录用户信息

❶ 新建Web窗体4_8a.aspx,页面上包含文本框TextBox1和TextBox2、两个标签、一个命令按钮btnConfirm。

❷ btnConfirm中输入以下代码:

```
protected void btnConfirm_Click(object sender, EventArgs e)
{
    Session["username"] = TextBox1.Text.Trim();
    Session["pwd"] = TextBox2.Text.Trim();
    Server.Transfer("4_8b.aspx");
}
```

说明:将新对象增加到Session集合中的方法为:

```
string str1 = "慕容复";
Session["var1"] = str1;
```

也可以为:

```
Session.Add("var1", str1)
```

❸ 新建 Web 窗体 4_8b.aspx,在 Page_Load 中输入以下代码:

```
protected void Page_Load(object sender, EventArgs e)
{
    string myStr = "用户名:" + Session["userName"] + "<br>" + "密码:" + Session["pwd"];
    Response.Write(myStr);
}
```

❹ 运行 4_8a.aspx,单击"确认"按钮。

在一次会话中 Session 对象中存储的值都是有效的。本例在 4_8a.aspx 页面中保存的值,到另一个页面 4_8b.aspx 中也可以使用这些值。浏览器在运行 4_8a.aspx 后,再运行一次浏览器(不是用当前运行后浏览器新打开一个页面,相当于打开一次新的会话),在地址栏中输入 localhost://.../4_8b.aspx,页面 4_8a.aspx 存储的会话变量是不存在的。

表 4-9 和表 4-10 列出了 Session 对象的属性和方法。

表 4-9 Session 对象的属性

属 性	描 述
Contents	获取对当前会话状态对象的引用
Count	获取会话状态集合的项数
IsCookieLess	获取一个值,该值指示 SessionID 是嵌入在 URL 中还是存储在 HTTPCookie 中
IsNewSession	获取一个值,该值指示是否与当前请求一同创建
SessionID	用来标识一个 Session 对象
TimeOut	获取并设置在会话状态提供程序终止会话之前各请求之间所允许的时间(以分钟为单位)

表 4-10 Session 对象的方法

方 法	描 述
Add	将新的项添加到 Session 集合中
Clear	从 Session 集合中清除所有对象,但不结束会话
Abandon	强行结束会话,并清除会话中的所有信息
CopyTo	将 Session 集合复制到一维数组中
Remove	删除会话状态集合中的项
RemoveAll	从会话状态集合中移除所有的键和值
RemoveAt	删除会话状态集合中指定索引处的项

4.5.4 Application

Application 对象是当前 Web 请求获取的 HttpApplicationState 对象,它启用 ASP.NET 应用程序中多个会话和请求之间的全局信息共享。一旦网站服务器被打开,就创建了 Application 对象,所有的用户共用一个 Application 对象并可以对其进行修改。Application 对象的这一特性使得网站设计者可以方便地创建聊天室和网站计数器等常用的 Web 应用程序。

Application 对象是一个对象集合,可以看作存储信息的容器,为所有用户共享。由于应用程序的状态变量可以同时被多个线程访问,为了避免产生无效数据,在设置状态变量值前,必须先锁定应用程序的状态,待数据值写完后,再解除锁定状态以供其他人员写入请求使用。以下代码是将 RequestCount 变量值加 1 的过程:

```
Application.Lock();                              //锁定
Application["RequestCount"] = ((int)Application["RequestCount"]) + 1;
Application.Unlock();                            //清除锁定
```

从应用程序状态中读取值,首先确定应用程序的变量是否存在,然后在访问该变量时将其转换为相应的类型。以下代码示例了检索应用程序的状态值 startTime,并将其转换为一个 DateTime 类型的名为 startTime 的变量:

```
if(Application["startTime"]!= null)
{
    DateTime startTime = (DateTime)Application["startTime"];
}
```

Application 对象的其他属性与方法基本上与 Session 的相同,此处不再逐一罗列。

例 4-9 使用 Application 对象编写一个简单的聊天程序,程序运行结果如图 4-11 所示。

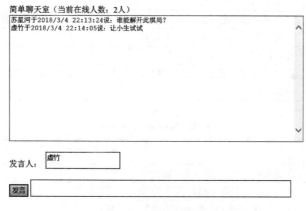

图 4-11 简单聊天室

❶ 新建 Web 窗体 4_9.aspx,网页上放置文本框 txtShowSay、txtSay、txtUser、命令按钮 btnSay、标签 label1,分别用于显示聊天内容、输入聊天内容、聊天人、聊天按钮、显示在线人数。

❷ 在该网页上编写以下事件过程:

```
protected void Application_Start(object sender, EventArgs e)
{
    Application["chart"] = "";
    Application["chatNum"] = null;
}
protected void Application_End(object sender, EventArgs e)
```

```
    {
        Application["chart"] = "";
        Application["chatNum"] = null;
    }
    protected void btnSay_Click(object sender, EventArgs e)
    {
        if (txtUser.Text == "")
        {
          Response.Write("<script>alert('必须输入姓名!')</script>");
          Response.End();
        }
        if (txtSay.Text == "")
        {
          Response.Write("<script>alert('必须输入要说的话!')</script>");
          Response.End();
        }
        int num;
        Application.Lock();
        if (Application["chatNum"] == null)
        {
            num = 0;
        }
        else
        {
            num = (int)Application["chatNum"];
        }
        num++;
        Application["chatNum"] = num;
        Application["chats"] += txtUser.Text + "于" + DateTime.Now.ToString() + "说:" + txtSay.Text + "\n";
        Application.UnLock();
        Label1.Text = "简单聊天室(当前在线人数:" + num + "人)";
        txtShowSay.Text = Application["chats"].ToString();
        txtSay.Text = "";
    }
```

❸ 运行该程序,输入"发言人"和"发言内容",然后单击"发言"按钮,用其他浏览器再次运行该程序,这样就可以实现两个人聊天了。

说明:

① Application_Start 和 Application_End 两个事件用于在应用程序启动和退出时执行。

② 本网页不能即时显示另一个网友的聊天信息,只有在单击"发言"按钮后才能显示所有网页的即时聊天信息。要实现刷新功能,可以借助 Ajax 完成。

习题与思考

(1) 在某个 Web 应用程序中,需要在 a.aspx 中将浏览器跳转到 www.163.com,可以使用的语句是(   )。

A. Response.Transfer("www.163.com");

B. Server.Transfer("www.163.com");

C. Response.Redirect("www.163.com");

D. Server.Redirect("www.163.com");

(2) 下面代码运行后浏览器显示的结果是(　　)。

```
protected void Page_Load(object sender, EventArgs e)
{
  Response.Buffer = true;
  Response.Write("欢迎你");
  Response.Flush();
  Response.Write("welcome you");
  Response.End();
  Response.Write("希望你喜欢该课程");
}
```

A. welcome you　　　　　　　　B. 希望你喜欢该课程

C. 欢迎你　　　　　　　　　　D. 欢迎你 welcome you

(3) 在 a.htm 中有一链接 click，单击后，希望将 a＝1、b＝2 传到下一个页面 a.aspx，a.htm 中链接的代码为(　　)。

A. ＜a href＝a.aspx? a＝1 & b＝2＞click＜/a＞

B. ＜a href＝a.aspx＋a＝1 and b＝2＞click＜/a＞

C. ＜a href＝a.aspx? a＝1&b＝2＞click＜/a＞

D. ＜a href＝＞click＜/a＞

(4) a.htm 中有以下代码：

```
< form action = "a.aspx" method = "post">
姓名< input type = "text" name = "txtName">
< input type = submit value = "ok">
</form>
```

单击"提交"按钮后，在 a.aspx 页面中需要得到 a.htm 页面中"姓名"后面文本框 txtName 的值，然后放在字符串变量 strName 中，使用的语句是(　　)。

A. strName＝Request.QueryString["txtName"];

B. strName＝Response.Form["txtName"];

C. strName＝Response.QueryString["txtName"];

D. strName＝Request.Form["txtName"];

(5) a.htm 中有以下代码：

```
< form action = "a.aspx" method = "get">
姓名< input type = "text" name = "txtName">
< input type = submit value = "ok">
</form>
```

单击"提交"按钮后，在 a.aspx 页面中需要得到 a.htm 页面中"姓名"后面文本框 txtName 的值，放在字符串变量 strName 中，使用的语句是(　　)。

A. strName=Request.QueryString["txtName"];

B. strName=Response.Form["txtName"];

C. strName=Response.QueryString["txtName"];

D. strName=Request.Form["txtName"];

(6) 下边的程序执行完毕后,页面上显示的内容是(　　)。

```
protected void Page_Load(object sender, EventArgs e)
{
  Response.Write("<a href = 'http://www.ustb.edu.cn'>北京科技大学</a>");
}
%>
```

A. 北京科技大学

B. <a href='http://www.ustb.edu.cn'>北京科技大学</a>

C. 北京科技大学(超链接)

D. 程序出现错误,无法输出

(7) 如果希望知道客户端的 IP 地址,可以使用 Request 对象的(　　)方法。

A. Servervariables　　　B. Form　　　C. Cookies　　　D. QueryString

第 5 章

C#语言基础知识

（1）类的定义和类的封装性。
（2）类的继承和多态性。
（3）数组和泛型。
（4）序列化和反序列化。

5.1 类

5.1.1 类的定义

在面向对象编程中，类是最基本的编程结构。类是由字段数据及操作这个数据的成员（如构造函数、属性、方法、事件等）构成的。下面建立一个 Person 类，该类具有 name 和 gender 两个属性以及 Insert()、Show()、Delete()3 个方法，通过这 3 个方法，向数据库中插入数据、显示和删除数据库中的记录。

在建立类前，可先用类图将类画在纸上：类名写在类图的最上面，各个属性和方法写在下面，这样一眼就能够看到该类的全部内容。Person 类的类图可表示为图 5-1。

❶ 在 students 数据库中，建立一个表 person，该表中有 name 和 gender 两个字段，字段类型都是字符。

❷ 新建类文件 Person.cs，引入以下命名空间：

using System.Data;
using System.Data.SqlClient;

❸ 在类中写入以下代码：

public class Person

图 5-1 Person 的类图

```csharp
{
    public string name;
    public string gender;
    public void Insert()
    {
        string cnnStr = "server = . ;database = students;uid = sa;pwd = 3500";
        SqlConnection cnn = new SqlConnection(cnnStr);
        string cmdStr = "insert into person(name,gender) values ('" + this.name + "','" + this.gender + "')";
        SqlCommand cmd = new SqlCommand(cmdStr,cnn);
        cnn.Open();
        cmd.ExecuteNonQuery();
        cnn.Close();
    }
    public void Show()
    {
        string cnnStr = "server = . ;database = students;uid = sa;pwd = 3500";
        SqlConnection cnn = new SqlConnection(cnnStr);
        string cmdStr = "select * from person";
        SqlCommand cmd = new SqlCommand(cmdStr, cnn);
        cnn.Open();
        SqlDataReader reader = cmd.ExecuteReader();
        while (reader.Read())
        {
            HttpContext.Current.Response.Write("姓名：" + reader["name"].ToString() + ",性别：" + reader["gender"].ToString() + "</br>");
        }
        cnn.Close();
    }
    public void Delete(string name)
    {
        string cnnStr = "server = . ;database = students;uid = sa;pwd = 3500";
        SqlConnection cnn = new SqlConnection(cnnStr);
        string cmdStr = "delete from person where name = '" + name + "'";
        SqlCommand cmd = new SqlCommand(cmdStr, cnn);
        cnn.Open();
        cmd.ExecuteNonQuery();
        cnn.Close();
    }
}
```

说明：

① C#中声明变量的格式是：变量作用域 变量类型 变量名，如 public string name;声明字段变量 name,类型是 string,访问范围是 public,表示 name 可以从对象及任何派生类中访问。C#中用";"表示一条语句的结束。

C#中的数据类型分为值类型和引用类型。值类型的变量内含变量值本身,值类型包括整数类型、字符类型、布尔类型、实数类型、结构体类型、枚举类型；引用类型的变量不直接存储所包含的值,它存储的是一个指针,该指针指向它要存储数据的另一块内存位置。引用类型包括类、委托、数组和接口。表 5-1 是 C#中内建的数据类型。

表 5-1 C♯内建系统类型

C♯简化符号	系统类型	范围	作用
bool	System.Boolean	true 或 false	表示真或假
sbyte	System.sByte	$-128\sim 127$	带符号的 8 位数
byte	Sytem.Byte	$0\sim 255$	无符号的 8 位数
short	System.Int16	$-32768\sim 32767$	带符号的 16 位数
ushort	System.UInt16	$0\sim 65535$	无符号的 16 位数
int	System.Int32	$-2147483648\sim 2147383673$	带符号的 32 位数
uint	System.UInt32	$0\sim 4294967295$	无符号的 32 位数
long	System.Int64	$-9223372036854775808\sim 9223372036854775807$	带符号的 64 位数
ulong	System.UInt64	$0\sim 18446744073709551615$	无符号的 64 位数
char	System.Char	$U+0000\sim U+ffff$	16 位的 Unicode 字符
float	System.Single	$-3.4\times 10^{38}\sim +3.4\times 10^{38}$	32 位浮点数
double	System.Double	$\pm 5.0\times 10^{-324}\sim \pm 1.7\times 10^{308}$	64 位浮点数
decimal	System.Decimal	$\pm 1.0\times 10^{-28}\sim \pm 7.9228\times 10^{28}$	128 位带符号数
string	System.String	受系统内存限制	一个 Unicode 字符集合
object	System.Object	任何类型都可以保存在 object 变量中	.NET 中所有类型的基类

从表 5-1 中可以看出，public decimal 与 public System.Decimal 是相同的，前者是 C♯专门的简写，后者是.NET 定义的系统类型。其他如 bool 与 Boolean、char 与 Char 的定义变量是相同的。

C♯的访问符除 public 外，还有 private、protected、interval、protected interval。本章只讲解 public 和 private。

② Person 类中定义了两个公共字段数据，即 name、gender。一般不会将字段数据设置为公共，而是通过类型属性对数据提供受控制的访问，在这里是为了论述简单。

③ Person 类可以具有某种行为，如增加新成员、列举成员、删除成员等，这里定义了 Insert()方法表示"增加"新成员；Show()方法表示显示全部成员、Delete()删除成员。Insert()、Show()、Delete()访问数据库，使用的是 ADO.NET，需要引入相应的命名空间。

④ string cnnStr = "server=.;database=students;uid=sa;pwd=3500";指定连接 SQL Server 的字符串。"."表示访问的是当地数据库，如果要访问的是网络数据库可改成 IP；database 指定要访问的数据库名，uid 指定用户名，pwd 指定用户密码。

⑤ 通过命令对象 cmd 执行 SQL。Insert()方法中 cmdStr 由于 name 和 gender 在 person 表中的数据类型是字符，故拼接字符串时变量两边要加上单引号。Show()方法中使用 DataReader 读取表中的记录。

ADO.NET 的详细用法见第 6、7 章。

5.1.2 使用类建立对象

Person 类的定义完成后，要将 Person 类使用 New 实例化为对象，如建立 p1 对象：

```
Person p1 = new Person();
```

也可以将上面的语句分成两句：

Person p1;
p1 = new Person();

必须通过 New 关键字把引用赋给对象以后，这个引用才能指向内存中有效的类实例。

新建 5-1.aspx 文件，从工具箱中拖曳 3 个命令到页面中，分别在 3 个命令按钮的 Click 事件中写入代码：

```
//增加成员的代码
Person p1 = new Person();
p1.name = "林大果";
p1.gender = "男";
p1.Insert();
p1.name = "林小果";
p1.gender = "女";
p1.Insert();
```

C#中使用//表示单行注释。上面的代码建立了 p1 对象，为该对象的 3 个字段赋值，然后调用 Insert()方法将对象的字段值保存到 person 表中。使用对象的属性（或字段）和方法，使用"."操作符，对属性赋值的格式为：对象名.属性=值。

```
//列出全部成员
Person p1 = new Person();
p1.Show();
//删除某成员如"林大果"
Person p1 = new Person();
p1.Delete("林大果");
```

上述 3 段代码中都有 Person p1 = new Person();类实例化，可以将该语句放在：

```
public partial class 类名
{
    Person p1 = new Person();
    protected void Page_Load(object sender, EventArgs e)
    {
    }
}
```

变量 p1 改变了作用域，可以在本页面的任何事件中使用，各命令按钮的事件中可以删除语句 Person p1 = new Person()。

运行 5-1.aspx 文件，单击"增加"按钮，再单击"输出"按钮，显示结果如图 5-2 所示。在 SQL Server 中浏览 person 表，表中增加了两条记录。

对象方法的使用格式为：对象名.方法名（传入的参数列表）。这里 Insert()方法中没有需要传入的参数。

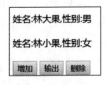

图 5-2　显示成员

1. 构造函数

如果在类实例化时，想直接为类中定义的字段或属性赋值。例如：

```
Person p1 = new Person("林大果", "男");
```

默认情况下，类会提供一个与类名相同的函数，称之为构造函数。默认的构造函数确保所有字段数据都设置为正确的默认值。如果对默认的赋值不满意，可以重新定义构造函数以满足需求。下面的代码更新了 Person 类中的构造函数：

```
public class Person
{
   public string name;
   public string gender;
   public Person()
   {
      name = "张三";
      gender = "男";
   }
   public void Insert()
   {
   }
      …
}
```

通过重新定义构造函数，强制所有的 Person 对象，其 name＝"张三"、gender＝"男"。在类中再定义一个构造函数：

```
public class Person
{
    public string name;
    public string gender;
    public Person()
    {
      name = "张三";
      gender = "男";
    public Person(string name, string gender)
    {
       this.name = name;
       this.gender = gender;
    }
    …
}
```

上面新增加的构造函数定义了 name、gender 两个传入参数，这些参数名并不要求与类中的字段名一定相同。如果相同，为了避免出现 name＝name，需要通过 this 以区分这两个 name。

上面同一个类中出现了两个构造函数。当同一个类中定义了同样的名称，但参数类型或者参数数量不同的方法时，称之为重载。有了上述的构造函数，新建 aspx 页面，运行以下代码：

```
protected void Page_Load(object sender, EventArgs e)
{
```

```
    Person p1 = new Person();
    Response.Write(p1.name);        //输出"张三"
    Person p2 = new Person("李梅","女");
    Response.Write(p2.name);        //输出"李梅"
}
```

2. 静态成员和静态类

在定义类成员、类方法时，如果加上 static 关键字，表示这些成员和方法属于类，而不是由类生成的对象。

```
public class Car
{
    static public string carName;
    public int speed;
}
//类实例化
Car myCar = new Car();
myCar.speed = 120;                  //语句正确
myCar.carName = "奥迪 A6";          //不正确,因 static 限定了 carName 属于类而非类的实例
Car.carName = "奥迪 A6";            //正确
```

声明类时，在类名前面加上 static，该类为静态类。静态类要求其内部成员必须都是静态的，以下代码编辑时会出现错误：

```
static public class Pet
{
    static public string petname;
    public int age;                 //语句编译时出错,改为: static public int age;
    public void Bark()              //语句编译时出错,改为: static public void Bark();
    {
        ...
    }
}
```

声明 Pet 为静态类，应该直接使用该类下的成员。例如：

```
string p1 = Pet.petname;
```

不能再使用 new 为静态类实例化：

```
Pet myPet = new Pet();              //语句错误
```

5.1.3 类的封装

类的封装核心是对象的内部数据不应该从对象实例直接访问。对象的数据应该被定义为私有的，如果调用者想改变对象的状态，需要间接使用公共成员。

前面定义 Person 类时：

```
class Person
{
```

```
    public string gender;
}
```

gender 表示性别,被定义为公共字段。其带来的问题是字段赋值时无法得到有效控制,下面的代码是有效的,但性别不能是"东方不败"!

```
Person p1 = new Person();
p1.gender = "东方不败";
```

1. 类属性的定义

在定义类时,表示对象状态的类成员一般都不会被标记为公共的。封装提供了保护状态数据完整性的方法。下面重新定义 Person 类。

```
public class Person
{
    private string _name;
    private string _gender;
    //name 属性
    public string name
    {
        get { return _name; }
        set{_name = value;}
    }
    //gender 属性
    public string gender
    {
        get { return _gender; }
        set
        {
            if (value == "男" || value == "女")
              _gender = value;
            else
              HttpContext.Current.Response.Write("性别只能是'男'或'女'");
        }
    }
    ...
}
```

说明:

① name、gender 属性的数据类型必须与其对应的字段属性_name 和_gender 的相同。

② C♯属性由作用域中定义的 get 作用域(访问方法)和 set 作用域(修改方法)构成。set 用于设置字段的属性值,value 表示调用者设置属性时传入的值;如果忽略 set 语句块,该属性变为只读属性,不能修改。

③ get 用于获取字段的属性值;如果忽略 get 语句块,该属性变为只写属性。

④ ||是"或"逻辑运行符,"与"逻辑运算符是 &&。

⑤ if 是分支语句,语法格式如下:

```
if(判断条件为真时)
{...}       //当一条语句时,可以省略{}
```

```
else
{...}
```

如果判断出现多路情况,虽然在 else 中可以嵌套 if,但书写比较麻烦,在判断是"等于"的情况下,使用 switch 比较简单:

```
switch(判断表达式)
{
  case:值 1
  {
   语句;
   break;
  }
  case:值 2
  {
   语句;
   break;
  }
  ...
  case:值 n
  {
   语句;
   break;
  }
  default:
  {
   ...
  }
}
```

2. 自动属性的定义

在创建属性封装数据的过程中,有时只需要简单地获取或者设置属性值,不需要对属性值做出业务逻辑上的判断,此种情况下,Person 类可以简化为:

```
public class Person
{
  public string name
  {get; set; }
  public string gender
  {get; set; }
  ...
}
```

3. 分部类型

当新建一个扩展名是 aspx.cs 的网页文件时,VS 会自动生成以下类:

```
public partial class 类名: System.Web.UI.Page
{
}
```

其中的 partial 关键字就是 C#中的分部类型。一个类如果有成千上万行的代码,导致 *.cs 文件非常长,该类也有可能由多人编写完成。此种情况下可以将一个类分布到多个 C#文件中。下面建立一个 aPerson 分部类做成两个分部类,分别保存在 PersonA.cs 和 PersonB.cs 两个文件中。

```
//PersonA.cs
public partial class aPerson
{
}
//PersonB.cs
public partial class aPerson
{
}
```

使用分部类时,要求分部类的每个部分一定要用关键字 partial。

5.1.4 类的继承

现在需要定义一个 Students 类,其属性在 Person 类属性的基础上,新增加 student_no (学号)和 faculty(系),如果要使 Students 类具有 Person 类的属性和方法,可以采用类的继承,Students 的类图表示如图 5-3 所示。

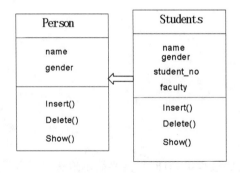

图 5-3 Students 类继承 Person 类

```
public class Students:Person
{
}
```

":"表示继承,Students 类继承 Person,Person 是基类或者父类。虽然 Students 类没有定义任何成员,但是 Students 拥有定义在父类 Person 中的每一个公共成员。

下面为 Students 类增加两个新的属性,即 student_no 和 faculty,并且增加构造函数:

```
public class Students:Person
{
  public string student_no
  {get; set; }
  public string faculty
  {get; set; }
  public Students(string student_no, string name, string gender, string faculty)
```

```
        {
            this.student_no = student_no;
            this.name = name;
            this.gender = gender;
            this.faculty = faculty;
        }
        public Students()
        {
        }
}
```

与 C++不同的是，C♯类只能有一个直接的基类，不能创建直接派生自两个或者两个以上基类的类型，即 C♯不支持多重继承，故以下 Students 类继承 Person 和 Family 两个类是错误的：

```
public class Students:Person,Family
{
    …
}
```

如果 Students 类不想被再往下继承，可以使用 sealed 关键字标记该类：

```
public sealed class Students:Person
{
    …
}
```

5.1.5 类的多态性

子类需要继承基类的所有方法、属性和字段时，继承相对比较简单。但往往是子类在继承基类后，需要修改基类的属性和方法。图 5-4 是一个 Bird 及其派生出的 Pigeon 和 Penguin 的类图。

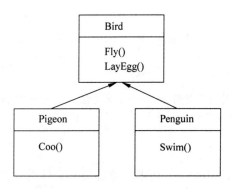

图 5-4 Pigeon 和 Penguin 继承 Bird

```
class Bird
{
    public void Fly()
    {
```

```
        //使鸟飞的代码
    }
    public void LayEgg()
    {
        //使鸟下蛋的代码
    }
}
class Pigeon:Bird
{
    public void Coo()
    {...}
}
class Penguin:Bird
{
    public void Swim()
    {...}
}
```

Bird 类中有 Fly()等方法,Pigeon 继承 Bird,Pigeon 会飞、会下蛋没有问题,但 Penguin 继承 Bird 类,Penguin 也变得会飞。子类 Penguin 继承父类 Bird 后,子类如何才能修改其从父类中继承下来的 Fly()?

(1) 父类中准备让子类修改的方法上增加 virtual 关键字:

```
class Bird
{
    //向 Fly()方法增加 virtual,表明允许子类修改该方法
    public virtual void Fly()
    {
        //使鸟飞的代码
    }
    public void LayEgg()
    {}
}
```

(2) 子类中在要重写父类的方法前,使用 override 关键字:

```
class Penguin:Bird
{
    //为了覆盖父类 Fly(),向子类增加一个同名的方法,而且要使用 override 关键字
    public override void Fly()
    {...}
}
```

Person 类中 Insert()、Delete()方法都是通过文本文件的操作实现的,Students 类继承了 Person 类,这两个方法也是对文本文件的操作。如果 Students 类中需要通过数据库的操作重写这两个方法,基类 Person 在定义这两个方法时,需要使用 virtual 关键字,而在子类 Students 中重写这两个方法时,需要使用 override 关键字。下面是 Person 类和 Students 类中这两个方法的重新定义:

```
public class Person
```

```
    {
        …
        //Insert 方法加上 virtual 后可以由派生类重写
        public virtual void Insert()
        {
            …
        }
        //Delete 方法加上 virtual 后可以由派生类重写
        public virtual void Delete(string name)
        {
            …
        }
    }
    public class Students:Person
    {
        …
        //派生类重写父类中的 Insert()方法
        public override void Insert()
        {
            …
        }
        //派生类重写父类中的 Delete ()方法
        public override void Delete(string name)
        {
            …
        }
    }
```

多态为子类提供了一种方式,使其可以重新定义由基类定义的方法,这个过程称为方法重写。父类中 virtual 关键字标记的方法称为虚方法,子类中要改变方法的实现细节,就必须用 override 关键字。

5.2 集合与泛型

数组是保存集合类数据的一个好方法。C#定义数组的方法如下:

```
//创建一个包含 3 个元素的字符串数组,编号为 0、1、2,并且初始化
string[] strFood = {"egg","noodle","fish" };
//创建一个包含 4 个元素的整型数组,编号为 0,1,2,3
int[] intArray = new int[4];
```

定义 intArray 后就可以逐个填充数组中的各个元素:

```
intArray[0] = 20;
intArray[1] = 21;
intArray[2] = 23;
```

要输出数组 intArray 的每一个值,可以:

```
foreach(int i in intArray)
```

```
{
    Response.Write(i + "<br>");
}
```

说明：

① intArray 包含 4 个元素，但只给 3 个元素赋值，intArray[4]默认值是 0。

② foreach 循环在遍历数组时非常方便。

intArray 数组的大小是固定的，如果：

```
intArray[4] = 50;
```

由于 intArray 声明只包含 4 个元素，故运行时将出现错误。在多数情况下，需要一种更加灵活的数据结构，希望容器可以动态伸缩，或者只保存满足某个标准的对象。为了摆脱上述简单数组的限制，.NET 基础类库发布了很多包含集合类的命名空间，可以分为非泛型集合和泛型集合两大类。

非泛型集合：命名空间是 System.Collections，常用的类有 ArrayList、HashTable、Queue、SortedList、Stack。

泛型集合：命名空间是 System.Collections.Generic，常用的类型有 List<T>、Queue<T>、Stack<T>、Dictionary<Key,Value>等，其中 T 表示某种数据类型，Key 表示键，Value 表示值。

需要说明的是，目前大多数.NET 项目都不会使用非泛型集合，而是使用泛型集合，原因是非泛型集合类型不安全，操作数值类型的数据时，CLR（公共语言运行时）必须执行大量的内存转换操作，从而降低程序的运行效率。本书对此不作详细讨论，而是直接介绍泛型集合。

5.2.1 泛型集合 List<T>的使用

List<T>是 System.Collections.Generic 命名空间中最常使用的类型，其可以动态地改变大小，T 表示类型参数，可以理解为占位符，占位符可以是表 5-1 所示的 int、string、float 等类型，也可以是自己定义的类名称。使用方法如：

```
List<int> intArray = new List<int>();
intArray.Add(10);
intArray.Add(15);
```

也可以在定义 intArray 时对其初始化：

```
List<int> intArray = new List<int>(){10,15,20};
```

例 5-2 使用泛型建立 Students 对象的列表。

```
List<Students> StudentsList = new List<Students>()
{
    new Students{student_no = "20160205",name = "张颖",gender = "女",faculty = "资源系"},
    new Students{student_no = "20160216",name = "张明",gender = "男",faculty = "土木系"},
    new Students{student_no = "20160225",name = "赵东",gender = "男",faculty = "环境系"}
};
//插入一个新的 Students
```

```
StudentsList.Insert(1, new Students { student_no = "20160207", name = "刘超", gender = 
"男", faculty = "资源系" });
//增加一个新的Students
StudentsList.Add(new Students { student_no = "20160230", name = "李娜", gender = "女",
faculty = "计算机系" });
Response.Write("一共有" + StudentsList.Count + "个学生<br>");
//列出所有StudentsList
foreach (Students s in StudentsList)
{
    string strStudent = string.Format("学号：{0},姓名：{1},性别：{2},系：{3}",s.student_no,s.name,s.gender,s.faculty);
    Response.Write(strStudent + "<br>");
}
```

运行结果如图 5-5 所示。

```
一共有5个学生
学号：20160205，姓名：张颖，性别：女，系：资源系
学号：20160207，姓名：刘超，性别：男，系：资源系
学号：20160216，姓名：张明，性别：男，系：土木系
学号：20160225，姓名：赵东，性别：男，系：环境系
学号：20160230，姓名：李娜，性别：女，系：计算机系
```

图 5-5 例 5-2 运行结果

说明：

① 通过初始化，将对象填充到 List<T> 中，相当于多次执行 StudentsList.Add()。

② StudentsList.Insert(1,…) 向 StudentsList 中插入记录，1 表示插入的索引位置。

③ StudentsList.Count 得到 StudentsList 对象的总数。

④ List<T> 有一个方法 ToArray()，可以返回一个对象数组：

```
Students[] ArrayOfStudent = StudentsList.ToArray();
//列出ArrayOfStudent的name
for(int i = 0;i < ArrayOfStudent.Length;i++)
{
    Response.Write(ArrayOfStudent[i].name);
}
```

5.2.2 泛型集合 Dictionary<Key,Value> 的使用

Dictionary 提供快速的元素查找，其结构是 Dictionary<Key,Value>，它包含在 System.Collections.Generic 命名空间中。在使用前，必须声明它的键类型和值类型。下面以 Key 的类型为 int，Value 的类型为 string 为例，说明其用法。

（1）创建及初始化：

```
Dictionary<int,string> myDictionary = new Dictionary<int,string>();
```

（2）添加元素：

```
myDictionary.Add(1,"C#");
myDictionary.Add(2,"C++");
myDictionary.Add(3,"ASP.NET");
```

```
myDictionary.Add(4,"MVC");
```

(3) 通过 Key 查找元素：

```
if(myDictionary.ContainsKey(2))
{
    string result = string.Format("Key:{0},Value:{1}", 2, myDictionary[2]);
    Response.Write(result);
}
```

(4) 遍历键值对：

```
foreach (KeyValuePair< int, string > kvp in myDictionary)
{
    string result = string.Format("Key:{0},Value:{1}", kvp.Key, kvp.Value);
    Response.Write(result + "< br >");
}
```

(5) 获得值和键的集合：

```
//获取值的集合
foreach (string s in myDictionary.Values)
{
    string result = string.Format("value = {0}", s);
    Response.Write(result + "< br >");
}
//获取键的集合
foreach (int s in myDictionary.Keys)
{
    string result = string.Format("key = {0}", s);
    Response.Write(result + "< br >");
}
```

(6) 统计字典中的键值对数：

```
int howMany = myDictionary.Count;
```

5.3 其他数据类型

5.3.1 DateTime 和 TimeSpan

DateTime 类型包括表示某个日期（年、月、日）的数据及时间值表示值的范围在公元 0001 年 1 月 1 日午夜 12：00：00 到公元 9999 年 12 月 31 日午夜 11：59：59 之间的日期和时间，其位于 System 命名空间中。用户可以通过以下语法格式定义一个日期时间变量：

```
DatrTime 日期时间变量 = new DateTime(年,月,日,时,分,秒)
```

如：

```
DateTime dt1 = new DateTime(2018,1,20);
DateTime dt2 = new DateTime(2018,10,30,8,15,20);
```

其中,dt1 是 2018 年 1 月 20 日零点零分零秒;dt2 是 2018 年 10 月 30 日 8 点 15 分 20 秒。

DateTime 常用的属性和方法见表 5-2 和表 5-3。

表 5-2　DateTime 常用的属性

属　性	说　明	示　例
Date	获取日期	
DayOfWeek	获取某日期为星期几	
DayOfYear	获取某日期为该年中的第几天	DateTime dt1 = new DateTime(2018, 11, 20); int day = dt1.DayOfYear;　　//day 为 324
Year	获取某日期的年	
Month	获取某日期的月	
Day	获取某日期的天	
Hour	获取某日期的小时部分	
Minute	获取某日期的分钟部分	
Second	获取某日期的秒部分	
Now	获取一个 DateTime 对象在计算机上的当前日期和时间	DateTime dt1 = DateTime.Now;
TimeOfDay	获取某日期当天的时间	
Today	获取当前日期	

表 5-3　DateTime 常用的方法

方　法	说　明	方法类型
AddYears	为某日期加上指定的年份数	非静态方法
AddMonths	为某日期加上指定的月份数	
AddDays	为某日期加上指定的天数	
AddHours	为某日期加上指定的小时	
AddMinutes	为某日期加上指定的分钟数	
AddSeconds	为某日期加上指定的秒数	
AddMillseconds	为某日期加上指定的毫秒数	
CompareTo	将某日期与指定的对象或者值类型进行比较,返回一个相对值的指示	
DaysInMonth	返回指定年月中的天数	静态方法
IsLeapYear	返回指定年月是否是闰月的指示	
Parse	将表示日期和时间的字符串转换为 DateTime 类型,如: string tt = "2018,10,20"; DateTime dt2 = DateTime.Parse(tt);	

TimeSpan 实例用来表示一个时间段的实例,两个时间的差就可以构成一个 TimeSpan 实例。

```
TimeSpan ts = new TimeSpan(4,20,30);        //构造函数接受时、分、秒
```

ts 减去 30 分钟 40 秒,结果是 3 时 49 分 50 秒:

Response.Write(ts.Subtract(new TimeSpan(0,30,40)).ToString());

注意:一个类的方法有静态方法和非静态方法之分。对于静态方法,只能通过类名来调用,而对于非静态方法,则需要通过类的对象来调用。

5.3.2 Convert 类

Convert 类位于 System 命名空间,用于将一个值类型转变成另一个值类型,其常用的方法见表 5-4,所有这些方法都是静态方法,可通过 Convert.方法名(参数)来使用。

表 5-4 Convert 类的常用方法

方 法	说 明
ToBoolean	将数据转换为 Boolean 类型
ToDateTime	将数据转换为 DateTime 类型
ToInt16	将数据转换为 16 位整数类型
ToInt32	将数据转换为 32 位整数类型
ToInt64	将数据转换为 64 位整数类型
ToNumber	将数据转换为 Double 类型
ToObject	将数据转换为 Object 类型
ToString	将数据转换为 String 类型

5.3.3 String 类

String 类位于 System 命名空间中,提供了返回字符串长度、查找当前字符串中子字符串、转换大小写等方法。String 类常用的属性和方法见表 5-5。

表 5-5 String 类常用的属性和方法

属性/方法	说 明
Length	这个静态属性返回当前字符串的长度
Compare()	用静态方法比较两个字符串。string.compare(string A,string B)的返回值是 A 与 B 的排序顺序,返回值是一个整型 int。当值是-1 时,A 在 B 的前面;当值是 0 时,A 和 B 在相同位置;当值是 1 时,B 在 A 的前面
Contains()	这个非静态方法判定当前字符串是否包含一个指定的子字符串
Equals()	这个非静态方法测试两个字符串对象是否包含同样的字符数据
Format()	此静态方法格式化一个字符串,用法: string today = DateTime.Now.ToString("yyyy 年 MM 月 dd 日"); Response.Write(String.Format("今天是:{0}",today));
Insert()	这个非静态方法将一个字符串插入到指定的字符串中
PadLeft()	这个非静态方法在字符串左边填充指定数量的字符
PadRight()	这个非静态方法在字符串右边填充指定数量的字符
Remove()	这个非静态方法从字符串中移去指定数量的字符
Replace()	这个非静态方法用新的字符替换字符串中指定的字符

续表

属性/方法	说明
Split()	这个非静态方法用指定的字符分隔某个字符串,分隔后的结果返回到数组中
Substring()	这个非静态方法从字符串中截取指定长度的子字符串。例如: string ss = "abcdefg"; Response.Write(ss.Substring(2, 5)); //输出 cdefg;2 是开始位置,5 是截取长度
ToUpper()	这个非静态方法用指定的字符串转换成大写
ToLower()	这个非静态方法用指定的字符串转换成小写
Trim()	这个非静态方法从字符串头部和尾部移除所有的指定字符

与 C 语言一样,在 C♯字符串中可以包含各种转义字符,用来限定字符数据被输出到输出流中。每个转义字符都以一个反斜线开始,后面跟一个特殊的标记,表 5-6 列出了最常见的转义字符。

表 5-6　字符串中使用的转义字符

字符	作用
\'	将一个单引号加入到一个字符串中
\"	将一个双引号加入到一个字符串中
\\	将一个反斜线加入到一个字符串中
\n	换行(Windows 平台)
\r	回车
\t	将一个水平制表符加入到一个字符串中

5.3.4　System.Text.StringBuilder 类

字符串一旦创建就不可修改大小,所以,如果对字符串添加或删除操作比较频繁,使用 String 类会很低效,.NET 类库在 System.Text 命名空间中提供了一个 StringBuilder 的类,与 System.String 类相似但在内存分配上更加高效。下面是 StringBuilder 输出一个表格的简单用法:

```
StringBuilder sb = new StringBuilder();
sb.AppendLine("<table><tr><td>汽车品牌</td></tr>");
sb.Append("<tr><td>");
sb.Append("奔驰");
sb.Append("</td></tr></table>");
Response.Write(sb);
```

可以看出,如果要按照指定的格式输出数据库中的数据,使用 StringBuilder 可以很方便地实现。

5.4　委托

委托是一种引用方法的类型,创建了委托,就可以声明委托变量,也就是委托实例化,实例化的委托就是委托对象,可以为委托对象分配方法,委托对象被分配方法后,委托对象将与该方法具有完全相同的行为。

下面通过一个简单的示例，说明委托的使用方法。

❶ 定义一个 delegateTest 类，类中有个乘法和加法的函数：

```
public class delegateTest
{
    public static double Multiply(double one, double two)
    {
        return one * two;
    }
    public static double add(double one, double two)
    {
        return one + two;
    }
}
```

❷ 新建 Web 窗体 5_8.aspx，在 Web 窗体中定义一个名为 JiSuan 的委托：

```
public delegate double JiSuan(double one, double two);
```

该委托包含两个 double 型的参数，可以看出委托的声明类似于函数，但不带函数体，且要使用 delegate 关键字。

❸ 在 Page_Load 事件中写入代码：

```
JiSuan js = delegateTest.Multiply;
Response.Write("20 * 30 = " + js(20, 30));
```

说明：JiSuan js = delegateTest.Multiply 定义委托的实例 js，将 Multiply 方法赋给该委托实例。js(20,30)调用委托。

委托类型派生自 Delegate 类，委托类型是密封的，不能从 Delegate 中派生出委托类型，也不能从中派生出自定义的类。由于实例化委托是一个对象，所以可以将其作为参数进行传递，也可以将其赋值给属性，这样，在方法中便可以将一个委托作为传递的参数来接受，并且可以调用该委托，称之为异步回调，是在较长的进程中完成后用来通知调用方的常用方法。在第 3 章的 3.8.4 小节中，$.get("3_8.ashx", sendData, function(data, status){})中的 function(data, status)就是回调函数。在 C# 中也有类似的用法，如：

```
public double JiSuanCallBack(double one, double two, JiSuan js)
{
    return js(one,two);
}
```

方法 JiSuanCallBack 包含一个参数 js，就是委托的实例。

5.5 JSON 的序列化和反序列

在使用 Ajax 向服务器请求信息时，服务器可以直接将对象通过序列化输出为 JSON 字符串，也可以通过反序列化将接收的 JSON 字符串生成对象。JSON 的序列化和反序列化有使用 JavaScriptSerializer 类、formatter 格式化器类、JSON.NET 类 3 种方式。

5.5.1　使用 JavaScriptSerializer 类序列化和反序列化

使用该类,需要引入命名空间:

using System.Web.Script.Serialization;

还需要在"解决方案资源管理器"中,右击"引用",选择快捷菜单中的"添加引用"命令,选择 System.Web.Extensions。

1. 序列化

建立 5_3.aspx 文件,引入命名空间 using System.Web.Script.Serialization;后,在 Page_Load()事件中输入以下代码:

```
List<Students> stu = new List<Students>();
stu.Add(new Students("9501","张三","男","计算机系"));
stu.Add(new Students("9601","李芳","女","环境系"));
stu.Add(new Students("9701","赵五","男","能源系"));
JavaScriptSerializer js = new JavaScriptSerializer();
string s = js.Serialize(stu);
Response.Write(s);
Response.End();
```

说明:

① 将 stu 序列化时,先定义一个 JavaScriptSerializer 对象 js,然后使用该对象的 Serialize()方法,就可以将 stu 序列化成 JSON 字符串。

② 在 aspx 文件中,在返回 Response.Write()后应该调用 Response.End()才能将数据写入到调用的页面,才能被 JQuery 的回调函数获取到返回的 JSON 数据。

③ 可以将 5_3.aspx 文件转换成为 5_3.ashx 一般处理程序文件,只是将 Response.Write(s);改为 context.Response.Write(s);,Response.End();可以省略。

运行 5_3.aspx,输出结果为:

[{"student_no":"9501","faculty":"计算机系","name":"张三","gender":"男"},
{"student_no":"9601","faculty":"环境系","name":"李芳","gender":"女"},
{"student_no":"9701","faculty":"能源系","name":"赵五","gender":"男"}]

下面利用 jQuery 将服务器序列化的数据显示在客户端上。

新建 5_3.html 文件,内容如下:

```
<body>
  <script type="text/javascript" src="../Scripts/jquery.min.js"></script>
  <div id="result"></div>
  <button onclick="display()">显示</button>
  <script>
        function display() {
            $.get("5_3.aspx", function (returnData, status) {
                var jsonObj = eval('(' + returnData + ')');
                $.each(jsonObj, function (index, value) {
                    $("#result").append("学号" + value["student_no"] + "姓名:" + value["name"] + "<br/>");
```

```
                });
            });
        }
    </script>
</body>
```

运行 5_3.html,浏览器中显示结果见图 5-6。

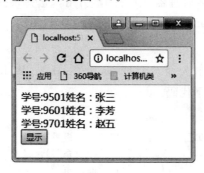

图 5-6 客户端解析 JSON 字符串后显示结果

说明:

① $.get()以 Ajax 方式向服务器端请求数据,用法见 3.5.4 小节。

② var jsonObj = eval('(' + returnData + ')');将 JSON 字符串转变为 JSON 对象。

③ $.each(jsonObj, function(index, value)是 jQuery 遍历对象的一种方法。jsonObj 必须是对象(数组、JSON、DOM 等)后台传过来的 JSON 字符串,要经过 eval、str.parseJSON()、JSON.parse(str)等方法转换为对象。index 是当前元素的位置,value 是值。

2. 反序列化

将 JSON 字符串转变为 JSON 对象。新建 5_4.aspx 窗体文件,在 Page_Load()事件中输入以下代码:

```
string desJson = "[{\"student_no\":\"9501\",\"name\":\"张三\",\"gender\":\"男\"},";
desJson += "{\"student_no\":\"9601\",\"name\":\"李芳\",\"gender\":\"女\"},";
desJson += "{\"student_no\":\"9701\",\"name\":\"赵五\",\"gender\":\"男\"}]";
JavaScriptSerializer js = new JavaScriptSerializer();
Students[] model = js.Deserialize<Students[]>(desJson);   //反序列化
for (int i = 0; i < model.Length; i++)
{
    string message = string.Format("学号={0},姓名={1},性别={2}", model[i].student_no, model[i].name, model[i].gender);
    Response.Write(message + "<br>");
}
```

说明:在使用 Deserialize 反序列化时,由于 desJson 字符串是个数组,故需要:

```
Students[] model = js.Deserialize<Students[]>(desJson);
```

或者

```
List<Students> model = js.Deserialize<List<Students>>(desJson);
```

但要将 for (int i = 0; i < model.Length; i++)改为

for (int i = 0; i < model.Count; i++)

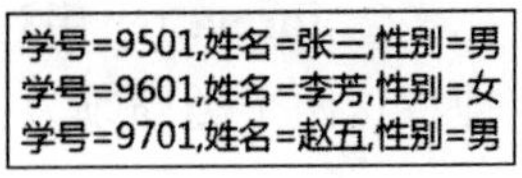

```
学号=9501,姓名=张三,性别=男
学号=9601,姓名=李芳,性别=女
学号=9701,姓名=赵五,性别=男
```

图 5-7 反序列化结果

而不能改为：

Students model = js.Deserialize<Students>(desJson);   //错误提示：数组的反序列化不支持类型

程序运行结果见图 5-7。

5.5.2 使用 formatter 格式化器序列化和反序列化

序列化工作由一个特定的格式化器(formatter)完成，每个格式化器都提供 Serialize 和 Deserialize 两个方法。当格式化器将某个对象序列化后，将结果放入一个流(Stream)中，因此可以包容任何序列化格式。一个对象被存储于一个流中，对象的状态可以被存储于磁盘上(或者说是 persistent：被持久化)。

需要引入命名空间：

```csharp
using System.Runtime.Serialization;
using System.Runtime.Serialization.Formatters.Binary;
using System.IO;
```

新建立一个 5_5.aspx，引入命名空间后，在 Page_Load()事件中输入以下代码，程序运行结果见图 5-7。

```csharp
List<Students> stu = new List<Students>();
stu.Add(new Students("9501","张三","男","资源系"));
stu.Add(new Students("9601","李芳","女","计算机系"));
stu.Add(new Students("9701","赵五","男","土木系"));
Stream stream = File.Open(Server.MapPath("DataFile.dat"), FileMode.Create, FileAccess.Write);
//序列化
BinaryFormatter formatter = new BinaryFormatter();
formatter.Serialize(stream, stu);
stream.Close();
//反序列化
Stream filestream = File.Open(Server.MapPath("DataFile.dat"), FileMode.Open);
//返回 List<student>,必须强制转换
List<Students> newStu = (List<Students>)formatter.Deserialize(filestream);
for (int i = 0; i < newStu.Count; i++)
{
    string message = string.Format("学号={0},姓名={1},性别={2}", newStu[i].student_no, newStu[i].name, newStu[i].gender);
    Response.Write(message + "<br>");
}
```

说明：

① 这种二进制序列化的方式将对象的状态保存在一个二进制文件(如 DataFile.dat)中，对网络传输很方便。

② formatter 序列化对象时,需要在 Person 类和 Students 类上标记[Serializable]特性,即将这两个类修改为:

```
namespace aspExample.chapter5
{
    [Serializable]
    public class Person
    {...
    }
    [Serializable]
    public sealed class Students:Person
    {...
    }
}
```

③ Stream stream = File.Open(Server.MapPath("DataFile.dat"), FileMode.Create, FileAccess.Write);利用 File 类提供的静态方法 Open,以流的形式存储在 DataFile.dat 文件中。Server 是 ASP.NET 提供的一个内部对象,Server.MapPath("DataFile.dat")表示建立的 DataFile.dat 文件,其位置与当前网页文件在同一文件夹下。

5.5.3 使用 Json.NET 序列化和反序列化

下载 Newtonsoft.Json.dll 后,在 VS 资源管理器中右击"引用",在弹出的快捷菜单中选择"添加引用"命令,将 Newtonsoft.Json.dll 引用到当前项目中,然后引入命名空间:

using Newtonsoft.Json;

下面的代码是使用 Json.NET 序列化和反序列化的示例。

```
List<Students> stu = new List<Students>();
stu.Add(new Students("9501", "张三", "男"));
stu.Add(new Students("9601", "李芳", "女"));
stu.Add(new Students("9701", "赵五", "男"));
//Json.NET 序列化
string jsonData = JsonConvert.SerializeObject(stu);
Response.Write(jsonData);
//反序列化
string desJson = "[{\"student_no\":\"9501\",\"name\":\"张三\",\"gender\":\"男\"},";
desJson += "{\"student_no\":\"9601\",\"name\":\"李芳\",\"gender\":\"女\"},";
desJson += "{\"student_no\":\"9701\",\"name\":\"赵五\",\"gender\":\"男\"}]";
List<Students> model = JsonConvert.DeserializeObject<List<Students>>(desJson)
for (int i = 0; i < model.Count; i++)
{
    string message = string.Format("学号={0},姓名={1},性别={2}", model[i].student_no, model[i].name, model[i].gender);
    Response.Write(message + "<br>");
}
```

习题与思考

(1) 为 Person 类增加一个属性 ID(表示身份证号码),要求该属性的限定条件是:
① ID 的长度必须是 18 位。
② gender 为"男"时,身份证倒数第 2 位数字必须是奇数;gender 为"女"时,身份证倒数第 2 位数字必须是偶数。

(2) 在 a.html 页面上解析来自 a.ash 中的 cars 序列化后的 JSON 字符串,并将其显示在页面上。对 cars 采用 JavaScriptSerializer 类序列化。a.html 页面运行结果如图 5-8 所示,请填空。

[1]品牌:Ford,颜色:red
[2]品牌:Jaguar,颜色:blue
[3]品牌:Mazda,颜色:green

图 5-8 页面运行结果

car 类的定义如下:

```
class car
{
    public string brand;
    public string color;
    public decimal displacement;
}
```

a.html 文件中代码:

```
<script type="text/javascript" src="../Scripts/jquery.min.js"></script>
<div id="result"></div>
    <script>
        $.get("_____", function (returnData, status) {
            var jsonObj = eval('(' + returnData + ')');
            $.each(jsonObj, function (index, value) {
                $("#result").append("[" + (index + 1) + "]品牌:" + _____ + ",颜色:" + _____ + "<br/>");
            });
        });
    </script>
```

a.ashx 文件中代码:

```
using System;
using System.Collections.Generic;
using System.Linq;
using System.Web;
using System.Web.Script.Serialization;
public void ProcessRequest(HttpContext context)
{
    context.Response.ContentType = "text/plain";
    List<car> cars = _____();
    cars.Add(new car {brand = "Ford", color = "red", displacement = 3.0M });
    cars.Add(new car {brand = "Jaguar", color = "blue", displacement = 2.0M });
    cars.Add(new car {brand = "Mazda", color = "green", displacement = 1.6M });
    _____;
    string s = js.Serialize(cars);
```

_____;
}

(3) 在 .NET 中使用 Newtonsoft.Json，解析下面 json 字符串，输出 Name 的值：

```
{
    Result:[{
    "Id":"2001",
    "Name":"程序员",
    "createTime":"2010-03-17 11:39:03",
    "BeginTime":"2010-03-17 00:00:00",
    "EndTime":"2010-03-31 00:00:00"
    }],
    Size:1,
    Page:1,
    IsSuccess:true
}
```

(4) 阅读程序填空。

定义 class A 和 class B 分别如下：

```
public class A
{
    public _____ string m1()
    {
        return "这是类 A 中的 m1";
    }
    public  string m2()
    {
        return "这是类 A 中的 m2";
    }
}
public class B : A
{
    public _____ string m1()
    {
        return "B类重写了 A 类中的 m1";
    }
}
```

在 Web 窗体中执行以下代码：

```
A a = new A();
B b = new B();
Response.Write(a.m1() + "<br>");
Response.Write(b.m1());
```

输出结果如图 5-9 所示。

(5) 阅读程序填空。

建立一个 person 类：

public class person

图 5-9 输出结果

```csharp
{
    public person()
    {
        _id = "110108197005202579";
    }
    private string _id;
    private string _name;
    public string ID
    {
        set {
            if (value.Length == 18)
            {
                _id = value;
            }
            else
            {
                _id = "身份证必须18位!";
            }
        }
        get { return _id; }
    }
    public string name
    {
        set { _name = value; }
        get { return _name; }
    }
    public string gender
    {
        get {
            if (Convert.ToInt16(_id.Substring(16, 1)) % 2 == 0)
            { return "女"; }
            else
            { return "男"; }
        }
    }
}
```

① 执行以下程序代码,运行结果是：_____

```csharp
person personA = new person();
personA.ID = "11010920010209";
Response.Write(personA.ID);
```

② 执行以下程序代码,运行结果是：_____

```csharp
person personA = new person();
personA.ID = "110109200102092368";
Response.Write(personA.gender);
```

③ 执行以下程序代码,运行结果是：_____

```csharp
person personA = new person();
```

Response.Write(personA.gender);

(6) 在自己的计算机上运行以下程序,并写出运行结果:

①

```
string s = "山羊上山山碰山羊脚";
string[] sArray = s.Split('羊');
Response.Write(sArray[1]);
```

运行结果是:＿＿＿＿＿＿＿＿

②

```
string bj = "北京科技大学";
bj = bj.Substring(0, bj.Length - 4);
Response.Write(bj);
```

运行结果是:＿＿＿＿＿＿＿＿

第 6 章

ADO.NET 连接和命令对象

（1）Connection 类。
（2）Command 类。
（3）构建可重用的访问数据库的代码。
（4）注入式 SQL 攻击的防范。
（5）事务。

6.1 ADO.NET 基础

.NET 框架带有自己的数据访问技术 ADO.NET，其能够根据不同的数据源，使用不同的数据提供程序模型。数据提供程序是连接数据源和应用程序间的桥梁，.NET 数据提供程序见图 6-1。

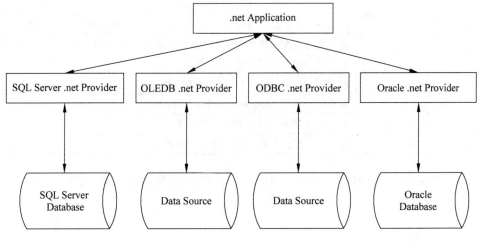

图 6-1 .NET 数据提供程序

不管是哪种数据提供程序,它们都有一系列类来提供核心功能,表 6-1 列举了一些核心公共对象、它们的基类(都定义在 System.Data.Common 命名空间)及其实现接口(都定义在 System.Data 命名空间)。

表 6-1　ADO.NET 常用的对象

对　象	基　类	实现的接口	说　明
Connection	DbConnection	IDbConnection	建立与断开数据源,提供事务对象的访问
Command	DbCommand	IDbCommand	对数据源执行所有 SQL 命令或存储过程
DataReader	DbDataReader	IDbDataReader	从数据源中读取只进且只读的数据流。所有 DataReader 的基类为 DbDataReader
Parameter	DbParameter	IDataParameter 和 IDbDataParameter	执行带参数的命令时表示的参数
DataAdapter	DbDataAdapter	IDbDataAdapter	用于存储 select、insert、update、delete 命令

.NET 虽然是针对不同的数据源提供了不同的数据库访问程序,但每个提供程序都基于同样的接口和基类,如 Connection 对象,都实现了 IDbConnection 接口,该接口定义了 Connection 对象的 Open()方法和 Close()方法,从而保证了不同数据源的 ADO.NET 对象具有相同的属性和方法。

ADO.NET 有两种类型的对象,即基于连接的对象和基于断开连接的对象。

(1) 基于连接的对象。它们是数据提供程序对象,包括 Connection、DataReader、Command 和 DataAdapter,它们连接到数据库,执行 SQL 语句,可以将查询的结果填充到 DataSet 中。命名空间与提供程序有关,如 SQL Server 的提供程序,其命名空间是 System.Data.SqlClient。

(2) 基于断开连接的对象。这些对象包括 DataSet、DataColumn、DataRow、DataRelation 等,它们与数据源独立,命名空间是 System.Data。

ADO.NET 的类保存在表 6-2 所示的命名空间中,每个提供程序都有自己的命名空间。

表 6-2　ADO.NET 的命名空间

命 名 空 间	说　明
System.Data	数据容器类,包括 DataSet、DataColumn、DataRow、DataRelation 等
System.Data.Common	包含大部分基本的抽象类,这些类实现了 System.Data 中的某些接口并定义了 ADO.NET 的核心功能
System.Data.OLEDb	包含用于访问 OLE DB 提供程序的类
System.Data.SqlClient	包含用于访问 SQL Server 提供程序的类
System.Data.OracleClient	包含用于访问 Oracle 提供程序的类
System.Data.ODBC	包含用于访问支持 ODBC 驱动程序的类

表 6-3 是使用 OLEDB 数据库操作组件和 SQL 数据库操作组件时,ADO.NET 各主要对象的写法。

表 6-3 不同数据源提供程序下 ADO.NET 主要对象的写法

数据库对象	OLE DB 数据库操作组件	SQL 数据库操作组件	Oracle 数据库操作组件
Connection	OleDbConnection	SqlConnection	OracleConnection
Command	OleDbCommand	SqlCommand	OracleCommand
DataAdapter	OleDbDataAdapter	SqlDataAdapter	OracleDataAdapter
DataReader	OleDbDataReader	SqlDataReader	OracleDataReader
CommandBuilder	OleDbCommandBuilder	SqlCommandBuilder	OracleCommandBuilder

6.2 Connection 对象

Connection 是 ADO.NET 最重要的对象,几乎和数据库关联的操作都要建立连接成功后才能进行,下面主要以访问 SQL Server 中的 students 库为例,说明 Connection 对象在使用时连接字符串的几种用法。

(1) 以 SQL Server 身份验证的方式登录时,如果采用 OLE DB 数据源提供程序,有以下代码:

```
string cnnString = "Provider = SqlOleDb;database = 数据库名;uid = 用户名;pwd = 密码;Server = servername";
OleDbConnection cnn = new OleDbConnection(cnnString);
```

Server:表示要访问的数据库服务器的 IP 地址。当访问当地 SQL Server 服务器时,可以将其值设置为".",127.0.0.1、SQL Server 服务器的名称或者 localhost。

uid:登录 SQL Server 的用户名。

pwd:登录 SQL Server 的密码。

database:要访问的数据库名。

(2) 以 SQL Server 身份验证的方式登录时,如果采用 SQL Server 数据源提供程序,有以下代码:

```
string cnnString = "server = servername;database = 数据库名; uid = 用户名;pwd = 密码";
SqlConnection Conn = new SqlConnection(cnnString);
```

连接字符串也可以写作:

```
string cnnString = "Data Source = servername;Initial Catalog = 数据库名; uid = 用户名;pwd = 密码"
```

在书写连接字符串时,连接字符串是用英文的分号(;)分隔的一系列名称/值对的选项,这些选项与顺序、大小写无关。

(3) 以 Windows 身份验证登录 SQL Server 时,有以下代码:

```
string cnnString = "Data Source = servername;Initial Catalog = 数据库名;trusted_connection = true";
```

或者

```
string cnnString = "server = servername;database = 数据库名; trusted_connection = true";
```

为了便于程序维护,一般是将数据库的连接字符串写在 web.config 文件的 <connectionString> 节中,代码如下:

```
<configuration>
<connectionStrings>
<add name = "SQLServerString" connectionString = "server = servername;database = students;
uid = sa;pwd = 3500" />
</connectionStrings>
</configuration>
```

程序在读取<connectionString>节内容前,需要做以下工作。
① 引入命名空间:

```
using System.Web.Configuration;
```

② 导入命名空间:打开"解决方案资源管理器"右击"引用",选择快捷菜单中的"添加引用"命令,在弹出对话框的"程序集"→"框架"下面选中 system.Configuration,两个步骤缺一不可。

使用以下代码,读取 web.config 中<connectionString>节 name 为 SQLServerString 的 connectionString 的值:

```
string cnnString = WebConfigurationManager.ConnectionStrings["SQLServerString"].ConnectionString;
```

本书默认已经将此连接串写入 web.config 文件中,以后连接数据库的字符串都以该种方式读取。

例 6-1 编程测试连接。
建立网页 6_1.aspx,新增加引入的命名空间:

```
using System.Data.SqlClient;
using System.Web.Configuration;
```

代码如下:

```
string cnnString = WebConfigurationManager.ConnectionStrings["SQLServerString"].ConnectionString;
SqlConnection conn = new SqlConnection(cnnString);
try
{
    conn.Open();
    Label1.Text = "数据库状态:" + conn.State;
}
catch (Exception err)
{
    Label1.Text = "数据库连接失败,错误原因:" + err.Message;
}
finally
{
    conn.Close();
    Label1.Text += ",<br/>数据库目前状态:" + conn.State;
}
```

由于连接消耗资源，连接要尽量晚打开，使用完毕后要尽早释放，上述代码即使在出现未处理的错误发生时，连接也能够在 finally 中关闭。也可以使用 using 语句声明一个正在使用的可释放的对象，将数据访问的代码放在 using 块中，using 语句一结束，CLR 会调用对象的 Dispose() 方法释放相应的对象。上述代码可改写为：

```
string cnnString = WebConfigurationManager.ConnectionStrings["SQLServerString"].ConnectionString;
SqlConnection conn = new SqlConnection(cnnString);
using(conn)
{
    conn.Open();
    Label1.Text = "数据库状态：" + conn.State;
}
Label1.Text += ",<br/>数据库目前状态：" + conn.State;
```

程序运行结果如图 6-2 所示。

图 6-2　连接运行结果

从运行结果可以看出，使用 using 块，块结束后 Connection 对象能够自动关闭。

6.3　Command 类

Command 类用于执行 SQL 语句。执行的 SQL 语句分为两种：一种是执行 select 语句，执行后获得查询结果；另一种是执行 insert、delete、update 及其他建立和修改库、表等的操作。Command 对象运行 select 语句后，使用 DataReader 将表中的记录读出来。

例 6-2　使用 Command 和 DataReader 读取 students 库中 student 表的全部记录（见 6_2.aspx）。

```
string cnnString = WebConfigurationManager.ConnectionStrings["SQLServerString"].ConnectionString;
SqlConnection conn = new SqlConnection(cnnString);
using (conn)
{
    SqlCommand cmd = new SqlCommand();
    cmd.CommandText = "select * from student";
    cmd.CommandType = System.Data.CommandType.Text;
    cmd.Connection = conn;
    conn.Open();
    using (SqlDataReader reader = cmd.ExecuteReader())
    {
        StringBuilder htmlStr = new StringBuilder();
```

```
        while (reader.Read())
        {
            htmlStr.Append("<li>");
            htmlStr.Append(reader["student_no"]);
            htmlStr.Append(reader["student_name"]);
            htmlStr.Append(reader["gender"]);
            htmlStr.Append("</li>");
        }
        Response.Write(htmlStr);
    }
}
```

说明：

① Command 对象最常用的属性是 CommandType、CommandText、Connection。CommandType 设置命令对象的类型，默认值是 CommandType.Text，命令对象执行的是一条 SQL 语句，还可以设置为一个存储过程或者一个表的名字。CommandText 与 CommandType 相对应，是 SQL 语句、表名或者存储过程名。Connection 设置命令使用的连接。

② 由于 CommandType 默认值是 CommandType.Text，示例中可以使用 Command 的一个构造方法，完成 Command 中 3 个属性的设置：

```
SqlCommand cmd = new SqlCommand("select * from student", conn);
```

如果要使用存储过程 GetStudent，也可以使用该构造方法，但需要指定 CommandType 的值为 CommandType.StoreProcedure，以便将表名和存储过程名相区分：

```
SqlCommand cmd = new SqlCommand("GetStudent", conn);
cmd.CommandType = System.Data.CommandType.StoreProcedure;
```

③ 设置 Command 对象属性后，Command 对象并没有运行。Command 对象提供了以下 3 个方法执行命令。

ExecuteReader()：执行 select 查询，并返回一个只读只进游标的 DataReader 对象。

ExecuteNonQuery()：执行非 select 的 SQL 语句。

ExecuteScalar()：执行 Select 查询并返回命令生成的记录集的第一行第一列的字段，常用于执行带有 Count()、Sum()等函数的 select 语句，返回统计结果。

④ 输出结果时采用了 StringBuilder，而没有直接使用"＋"操作符通过字符串连接的方式，对大的字符串而言，可以避免每次字符串连接时都要销毁和建立 String 对象，加快运行速度。

例 6-3 使用 ExecuteScalar 方法得到 students 表的记录总数。

```
string cnnString = WebConfigurationManager.ConnectionStrings["SQLServerString"].ConnectionString;
SqlConnection conn = new SqlConnection(cnnString);
using (conn)
{
    SqlCommand cmd = new SqlCommand("select count(*) from student",conn);
    conn.Open();
    int studentNum = (int) cmd.ExecuteScalar();
    Response.Write("共有学生：" + studentNum.ToString() + "人");
}
```

需要注意的是,由于 ExecuteScalar()返回一个对象,故需要将结果转换为别的数据类型。命令对象执行完毕后,要及时关闭连接对象。例 6-3 中由于使用了 using(conn),能够自动关闭连接,故没有使用 conn.Close()。

例 6-4 ExecuteNonQuery 方法的使用。

ExecuteNonQuery 方法执行不返回结果集的命令,如 delete、insert、update 等,返回的信息是受影响的行数,如果执行 create、alter 等其他 SQL 语句,返回 -1。以下代码在 students 库中创建一个新表 newStudent,创建 newStudent 前,先检查库中是否已经存在该表,如果存在,删除后再建立 newStudent。

```
string cnnString = WebConfigurationManager.ConnectionStrings["SQLServerString"].ConnectionString;
SqlConnection conn = new SqlConnection(cnnString);
using (conn)
{
    SqlCommand cmdA = new SqlCommand("select count(*) from sysobjects where id = object_id('newStudent')", conn);
    conn.Open();
    if ((int)cmdA.ExecuteScalar() != 0)
    {
        //newStudent 表库中已经存在
        using (SqlCommand cmdB = new SqlCommand("drop table newStudent",conn))
        {
            cmdB.ExecuteNonQuery();
        }
    }
    using (SqlCommand cmdB = new SqlCommand())
    {
        cmdB.CommandText = "create table newStudent(学号 char(8),姓名 char(10),性别 char(2))";
        cmdB.Connection = conn;
        cmdB.ExecuteNonQuery();
    }
    Response.Write("<script>alert('newStudent 表已经建立')</script>");
}
```

通过命令建立新表前先要检查新建立的表在库中是否已经存在。Sysobjects 位于 SQL Server 的"系统数据库"→master→"视图"→"系统视图"下,该视图有一列 id,通过设置查询条件 id=object_id('newStudent'),判断库中是否存在表 newStudent。SQL Server 提供 object_id()函数,查询数据库中是否存在指定名称的索引或者外键约束等。

6.4 DataReader 类

DataReader 以只读只进的方式从 select 查询的结果集中读取一条记录,它读取记录简单、快速,但不具有排序功能。**数据读取器只能处理 SQL 的 select 语句**,不能通过 insert、update、delete 请求来修改表中的记录。表 6-4 列出了 DataReader 的主要方法。

表 6-4　DataReader 的主要方法

方　法	说　明
Read()	将行游标前进到流的下一行。DataReader 新创建后,游标位置在第一行前,此时 Read()返回值是 false;如果行游标移到最后一行,则行游标返回值是 false
GetValue()	DataReader.GetValue(i)、DataReader[i]、DataReader["字段名"]含义相同,获得当前行第 i 列的值,其中 i 表示字段名在表中的序号,第 1 个字段序号为 0
GetName()	DataReader.GetName(i)获得第 i 列的字段名
NextResult()	当返回的 DataReader 包含多个数据集时,移动到下一个数据集
Close()	关闭 DataReader,当结束操作 DataReader 后要尽快关闭 DataReader,以释放该对象

例 6-2 使用 DataReader 读取记录时,可以使用 reader["字段名"]和 reader[字段序号]两种方式来取值,在不知道表中字段名的情况下,使用"字段序号"的方式更加方便。

例 6-5　在不知道表中字段名的情况下,使用 DataReader 显示表 course 中的全部记录。

```
string cnnString = WebConfigurationManager.ConnectionStrings["SQLServerString"].ConnectionString;
SqlConnection conn = new SqlConnection(cnnString);
using (conn)
{
    SqlCommand cmd = new SqlCommand();
    cmd.CommandText = "select * from course";
    cmd.CommandType = System.Data.CommandType.Text;
    cmd.Connection = conn;
    conn.Open();
    using (SqlDataReader reader = cmd.ExecuteReader())
    {
        while (reader.Read())
        {
            htmlStr.Append("<li>");
            for (int i = 0; i < reader.FieldCount; i++)
            {
                htmlStr.Append(reader[i].ToString() + " ");
            }
            htmlStr.Append("</li>");
        }
        Response.Write(htmlStr);
    }
}
```

例 6-6　使用 DataReader 获取多个结果集。

可以使用 DataReader 从一个命令对象获取多个结果集,如获取 student 和 course 表中的所有行,要写两个 select 语句,语句间用分号分隔,DataReader 从一个结果集转到另一个结果集,需要使用 DataReader 的 NextResult()方法,程序代码如下:

```
string cnnString = WebConfigurationManager.ConnectionStrings["SQLServerString"].ConnectionString;
SqlConnection conn = new SqlConnection(cnnString);
using (conn)
{
```

```
SqlCommand cmd = new SqlCommand();
cmd.CommandText = "select * from student;select * from course";
cmd.CommandType = System.Data.CommandType.Text;
cmd.Connection = conn;
conn.Open();
using (SqlDataReader reader = cmd.ExecuteReader())
{
    do
    {
        StringBuilder htmlStr = new StringBuilder();
        while (reader.Read())
        {
            htmlStr.Append("<li>");
            for (int i = 0; i < reader.FieldCount; i++)
            {
                htmlStr.Append(reader[i].ToString() + " ");
            }
            htmlStr.Append("</li>");
        }
        Response.Write(htmlStr);
    } while (reader.NextResult());
}
```

6.5 构建可重用的访问数据库的代码

程序开发中,一般不会将数据访问的代码与用户界面放在一起,而是将数据访问的逻辑单独存放到代码库中,目的是增加代码的重用性。三层架构提供了一种程序开发逻辑结构模式的一种分层重用机制,无论是 WebForm 开发模式还是 MVC(模型视图控制器)开发模式,都可以选择使用 3 层架构来布局程序开发的代码。在开发各种应用程序时,通常需要大量的代码,系统的"层"是相对代码的一种逻辑划分,并不一定是 3 层,如果系统复杂,可能是多层。分层的核心是外层一定不涉及任何的数据处理,它的任务就是设计界面、获取数据、输出数据。3 层架构一般分为数据访问层(Data Access Layer,DAL)、业务逻辑层(Business Logic Layer,BLL)和用户界面层(User Interface Layer,UIL)。本节以访问 students 库的 student 表为例,说明程序开发中多层架构的构建。

6.5.1 数据访问层

为了更方便地执行数据库的读写操作,一般根据数据库中表的结构先创建一个数据访问的类。本书中将访问数据库字段的类,全部放在 StudentDB.CS 文件中。该文件中定义了 studentData、scoreData、courseData 这 3 个类,用于访问 student 表、score 表和 course 表,代码如下:

```
#region student 的类
public class studentData
{
```

```csharp
private string _student_no;
private string _student_name;
private string _email;
private string _telephone;
private string _identification;
private int _id;
public studentData()
{
    //构造函数
}
public studentData(string student_no, string student_name, string email, string identification)
{
    //构造函数
    this.student_no = student_no;
    this.student_name = student_name;
    this.email = email;
    this.identification = identification;
}
public string student_no
{
    set { student_no = value; }
    get { return _student_no; }
}
public string student_name
{
    set { _student_name = value; }
    get { return _student_name; }
}
public string gender
{
    get
    {
        if (_identification.Length == 18)
        {
            if (Convert.ToInt16(_identification.Substring(16, 1)) % 2 == 0)
            { return "女"; }
            else
            { return "男"; }
        }
        else
        {
            { return "男"; }
        }
    }
}
public string email
{
    set { _email = value; }
    get { return _email; }
}
public string identification
```

```csharp
        {
            set
            {
                if (value.Length == 18)
                {
                    _identification = value;
                }
                else
                {
                    _identification = "身份证必须18位!";
                }
            }
            get { return _identification; }
        }
        public int id
        {
            get { return _id; }
        }
    }
    #endregion
    #region score 的类
    public class scoreData
    {
        public string student_no
        { set; get; }
        public string course_no
        { set; get; }
        public decimal score
        { set; get; }
        public int id
        {set;   get;}
    }
    public class courseData
    {
        public string course_no
        { set; get; }
        public string course_name
        { set; get; }
        public string course_address
        { set; get; }
        public string course_time
        { set; get; }
        public int teacher_ID
        { set; get; }
        public int id
        { set; get; }
    }
    #endregion
```

说明：

① studentData 类中 gender 属性是只读的，可以根据 identification 值判定出。

② scoreData 类、courseData 类的属性，采取自动属性建立的方法，与 studentData 中属性建立的不同之处在于 scoreData 中自动建立的属性不具有逻辑判断，如 studentData 中 identification 属性中对长度的判断。

studentData 类和 scoreData 类作为本书中默认使用的类。DAL 建立完成后，根据业务需求编写业务逻辑层。

6.5.2 建立连接的逻辑

新建 DbBase.CS 文件，在 DBbase 类中建立一个静态连接方法 ConnectionSqlServer 和关闭连接的方法 CloseSqlServer()（本章所有业务逻辑都放在 DBbase 类中）。

```
public class DBbase
{
    string cnnString = WebConfigurationManager.ConnectionStrings["SQLServerString"].ConnectionString;
    private static SqlConnection Conn = new SqlConnection(cnnString);
    public static bool ConnectionSqlServer()
    {
        try
        {
            Conn.Open();
            return true;
        }
        catch
        {
            return false;
        }
    }
    public static void CloseSqlServer()
    {
        if (conn.State == ConnectionState.Open)
        {
            conn.Close();
        }
    }
}
```

6.5.3 在 DBbase 类中建立查询数据的逻辑

利用前面介绍的 DataReader 可以从数据库中读取数据。下面介绍如何将读取出的数据返回给调用者：一种方法是将获得的数据存放到多维数组或者泛型；另一种是返回到 DataTable。

下面是使用泛型查询 student 表中的数据，建立一个 getAllStudents() 方法：

```
public static List<studentData> getAllStudents()
{
    List<studentData> student = new List<studentData>();
    string sql = "select student_no,student_name,email,identification from student";
    using (SqlCommand cmd = new SqlCommand(sql, conn))
```

```csharp
    {
        SqlDataReader reader = cmd.ExecuteReader();
        while (reader.Read())
        {
            student.Add(new studentData
            {
                student_no = reader["student_no"].ToString(),
                student_name = reader["student_name"].ToString(),
                email = reader["email"].ToString(),
                identification = reader["identification"].ToString()
            });
        }
        reader.Close();
    }
    return student;
}
```

上面的查询方法给出了查询 student 表中的全部记录。更多的情况是根据"学号"或者"姓名"查询指定的记录。下面是根据"姓名"查询的代码:

```csharp
public static List<studentData> getStudentsByName(string student_name)
{
    List<studentData> student = new List<studentData>();
    string sql = "select student_no,student_name,email,identification,id from student where student_name = '" + student_name + "'";
    using (SqlCommand cmd = new SqlCommand(sql, conn))
    {
        SqlDataReader reader = cmd.ExecuteReader();
        while (reader.Read())
        {
            student.Add(new studentData
            {
                student_no = reader["student_no"].ToString(),
                student_name = reader["student_name"].ToString(),
                email = reader["email"].ToString(),
                identification = reader["identification"].ToString(),
                id = (int)reader["id"]
            });
        }
        reader.Close();
        return student;
    }
}
```

说明:

① 字符串 sql 拼接时,由于 student 表中字段 student_name 的数据类型是字符型,故拼接变量时要在变量的两边加上单引号。对数值型字段,不需要单引号。对日期型字段,数据库不一样,界定符略有区别,如 SQL Server 是单引号、Access 是 #。

② 上面的查询代码中没有打开或关闭连接的代码,调用前需要单独执行打开或关闭连接的方法。

下面是使用泛型查询 course 课程表中的数据,建立一个 getAllCourse()方法:

```csharp
public static List<courseData> getAllCourse()
{
    List<courseData> course = new List<courseData>();
    string sql = "select * from course";
    using (SqlCommand cmd = new SqlCommand(sql, conn))
    {
        SqlDataReader reader = cmd.ExecuteReader();
        while (reader.Read())
        {
            course.Add(new courseData
            {
                course_no = reader["course_no"].ToString(),
                course_name = reader["course_name"].ToString(),
                course_address = reader["course_address"].ToString(),
                course_time = (string)reader["course_time"].ToString(),
                teacher_ID = reader["teacher_ID"].ToString(),
                id = (int)reader["id"]
            });
        }
        reader.Close();
    }
    return course;
}
```

下面建立一个 getScoreByCourseNo()方法,根据 course_no,查询出 student_no、student_name、score,查询的结果返回一个 DataTable:

```csharp
public static DataTable getScoreByCourseNo(string courseNo)
{
    string sql = "select student.student_no as student_no, student.student_name as student_name, score.score as score from score, student where student.student_no = score.student_no and score.course_no = '" + courseNo + "'";
    SqlDataAdapter adapter = new SqlDataAdapter(sql, conn);
    DataSet ds = new DataSet();
    adapter.Fill(ds);
    return ds.Tables[0];
}
```

6.5.4 在 DBbase 类中建立插入数据的逻辑

插入数据使用 SQL 的 insert 语句,通过命令对象的 ExecuteNonQuery 执行 insert 语句,完成数据的插入。

```csharp
public static void InsertStudent(studentData st)
{
    string sql = string.Format("insert into student(student_no, student_name," + ", gender, email, identification) values('{0}','{1}','{2}','{3}','{4}')", st.student_no, st.student_name, st.gender, st.email, st.identification);
    using (SqlCommand cmd = new SqlCommand(sql, conn))
```

```
        {
            cmd.ExecuteNonQuery();
        }
}
```

InsertStudent 方法中需要传入的参数类型为 studentData，也可重载该方法，传入多个相应的参数。由于要将"性别"的值保存到数据库中，重载时需要根据身份证编号判定出性别：

```
public static void InsertStudent(string student_no, string student_name, string email, string identification)
{
    string gender = "男";
    if (Convert.ToInt16(identification.Substring(17, 1)) % 2 == 0)
        { gender = "女"; }
    else
        { gender = "男"; }
    string sql = string.Format("insert into student(student_no, student_name," + "gender, email, identification) values('{0}','{1}','{2}','{3}','{4}')", student_no, student_name, email, identification);
    using (SqlCommand cmd = new SqlCommand(sql, conn))
    {
        cmd.ExecuteNonQuery();
    }
}
```

6.5.5　在 DBbase 类中建立更新数据的逻辑

更新记录可以使用 SQL 的 update 语句。更新记录时一般需要在 update 语句中指定更新条件，告诉计算机更新的是哪一条记录。为了更新方便，在建立数据库的每个表中，增加一个名为 id 的字段，将该字段的数据类型设置为 int 或者 bigint，将其"标识规范"属性的"是标识"设置为"是"，这样每增加一条记录，该字段值会自动增加。在显示数据的界面上隐藏 id，更新记录时，将选定更新记录的 id 传到该方法，完成记录的更新。

```
public static void UpdateStudent(int id, studentData st)
{
    string sql = string.Format("update student set student_no = '{0}', " + "student_name = '{1}', gender = '{2}', email = '{3}', identification = '{4}'," + "st.student_no, st.student_name, st.gender, st.email, st.identification)";
    using (SqlCommand cmd = new SqlCommand(sql, conn))
    {
        cmd.ExecuteNonQuery();
    }
}
```

6.5.6　界面层的设计

上述的数据访问层和业务逻辑层，可以应用于 ASP.NET、Windows 窗体、WPF（Windows Presentation Foundation）、控制台等应用程序。ASP.NET 中可以直接使用

Visual Studio 提供的控件,也可以使用其他框架如 jQuery EasyUI,建立用户界面。

下面直接使用 Visual Studio 控件,根据学生姓名,查询 student 表的内容,来简要说明界面层如何调用业务逻辑层。

例 6-7　根据输入学生的姓名,完成查询。

新建 aspx 文件,界面上放置一个标签、一个文本框和一个命令按钮,将以下代码放在命令按钮的单击事件中。程序运行结果见图 6-3。

```
protected void Button1_Click(object sender, EventArgs e)
{
    if (DBbase.ConnectionSqlServer() == true)    //打开连接
    {
        GridView1.DataSource = DBbase.getStudentsByName(TextBox1.Text.Trim());
        GridView1.DataBind();
        DBbase.CloseSqlServer();                 //断开连接
    }
    else
    {
        Response.Write("<script>alert('连接数据库发生错误')</script>");
    }
}
```

图 6-3　根据"姓名"查询的结果

说明:

① DBbase 类中的各种方法都是静态的,类无须实例化,可以直接使用。

② 设置 GridView1 的 DataSource 为 getStudentsByName,需要传入的参数是要 TextBox1 中输入的姓名,DataBind()方法用于绑定数据源。

③ 连接失败时,通过 ASP.NET 内部对象 Response 的 Write 方法,输出 JavaScript 代码,给出数据库连接失败的对话框提示。

6.6　SQL 注入攻击

6.5 节中都是以字符串拼接的方式,完成 SQL 语句中 insert、update、select 等的书写,如根据"姓名"查询,书写 SQL 的语句为:

```
string sql = "select student_no,student_name,email,identification from student where student_name = '" + student_name + "'";
```

图 6-3 中,在正常输入的情况下程序能够正常运行,但如果输入的内容中带有单引号如'a',执行查询时程序就会出现问题。通过对上述 sql 字符串的分析,发现将 sql 中 student_ name 替换为'a'时,sql 的值为:

select student_no,student_name,email,identification from student where student_name = ''a''

这样就出现了字符串拼接上的错误。如果访问者熟悉 SQL 语句,在查询的输入框中输入 a' or '1'='1,此时 sql 串变为:

select student_no,student_name,email,identification from student where student_name = 'a' or '1' = '1'

查询条件 student_name='a' or '1'='1'对任何记录都成立,查询结果就会显示出表中的全部记录,程序运行结果如图 6-4 所示,这样用户就非法获得了数据,网络安全受到威胁。如果是用户身份和密码验证,采用这种方式编程,攻击者可以轻易地绕开密码防护,进入到网络中。这种由于程序编码问题,程序开发者未预期地将 SQL 代码传入到应用程序的过程,就是 SQL 注入攻击。SQL 注入攻击是与数据库交互的 Web 应用程序面临的最严重的风险之一。

注入攻击可以有更加复杂的形式,如对 SQL Server,攻击者可以使用两个连接符"--"注释掉 SQL 语句的剩余部分;对 MySQL 使用"♯"、Oracle 使用";"。另外,攻击者还可以执行含有任意 SQL 语句的批处理命令。对于 SQL Server 攻击者只需要在新命令前加上";",攻击者可以采取这种方式删除其他表的内容,甚至调用 SQL Server 的系统存储过程 xp_cmdshell 命令行执行任意程序。在图 6-4 查询的文本框中,如果输入图 6-5 所示的内容,student 表中的数据将会被全部删除。

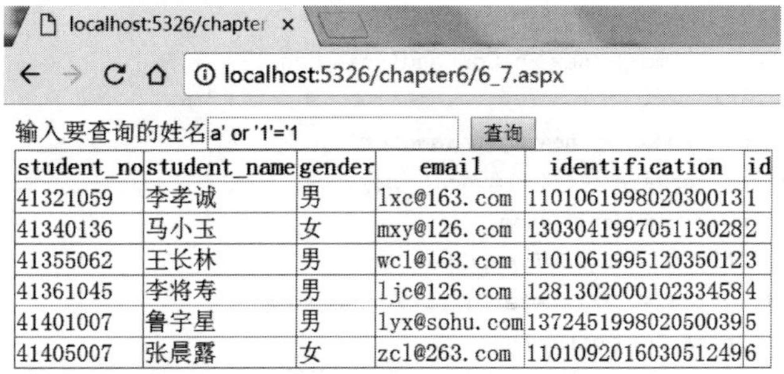

图 6-4 SQL 通过注入攻击非法获取数据

图 6-5 SQL 通过注入攻击非法删除表

注入 SQL 攻击对 Web 应用程序的威胁如此之大,需要采取如下防范措施。

(1) 限制用户文本框中输入内容的长度。通过设置文本框的 MaxLength 属性达到限制长度的目的。

(2) 对于文本框中输入的内容,禁止单引号、空格、--等的输入。通过 ASP.NET 验证控件如 RegularExpressionValidator,设置正则表达式,对特殊字符加以控制;如果文本框中一定要有特殊字符需要输入,此方法控制起来就比较麻烦。

(3) 最好的解决方法是使用参数化的命令或者是存储过程,防止注入 SQL 的攻击。

(4) 为用户设置数据库的访问权限,限制用户访问操作数据库或者执行扩展的系统存储过程。此方法能够预防删除表,但不能阻止攻击者偷看别人的信息。

6.7 参数化命令

参数化命令是在 SQL 文本中使用占位符的命令。占位符表示动态替换的值,它们可以通过 Command 对象的 Parameters 集合来传送。参数化命令执行速度快而安全,它只需要解析一次,而不像 SQL 语句那样每次被分配到 CommandText 时都要被解析。使用参数化命令的不足之处是代码稍微变长。

为了支持参数化查询,ADO.NET 的 Command 对象使用 Parameters 集合保存参数对象。该集合默认是空的,可以添加任意多的参数对象并映射到 SQL 语句中的占位符参数。SQL Server 中通过在 SQL 语句的参数前加 @ 符号,将 SQL 语句中的参数与 Command 对象参数集合中的某一成员相关联。

例 6-8 使用参数化命令对象,修改 6.5.3 小节根据"姓名"查询的代码。

```
public static List<studentData> getStudentsByNameA(string student_name)
{
    List<studentData> student = new List<studentData>();
    string sql = "select student_no,student_name,email,identification from student where student_name=@student_name";
    using (SqlCommand cmd = new SqlCommand(sql, conn))
    {
        SqlParameter param = new SqlParameter();
        param.ParameterName = "@student_name";
        param.Value = student_name;
        param.SqlDbType = SqlDbType.Char;
        param.Direction = ParameterDirection.Input;
        cmd.Parameters.Add(param);
        SqlDataReader reader = cmd.ExecuteReader();
        while (reader.Read())
        {
            student.Add(new studentData
            {
                student_no = reader["student_no"].ToString(),
                student_name = reader["student_name"].ToString(),
                email = reader["email"].ToString(),
                identification = reader["identification"].ToString(),
            });
        }
        reader.Close();
        return student;
    }
}
```

说明：

① sql 字符串中使用@student_name 作为占位符。

② SqlParameter param = new SqlParameter()定义一个参数对象 param，参数对象的 ParameterName 属性值要与 SQL 字符串中的占位符相对应。

③ 通过设置参数对象的 Value 属性，为参数赋值。cmd.Parameters.Add(param)将设置好的参数加入到参数集合中。

④ 参数对象的 Direction 属性指定参数是输入（Input）还是输出（Output），默认是 Input。

⑤ 如果参数有多个，需要不断定义和设置参数对象，这样程序写起来比较麻烦，代码中 6 条斜体语句可以使用以下的语句替代：

```
cmd.Parameters.AddWithValue("@student_name", student_name);
```

AddWithValue 的第 1 个参数是 SQL 语句中的占位符，第 2 个参数是替换点位符所采用的值。该方法可以自动识别参数的数据类型和大小，参数输入方向是 Input。

程序修改后，再次运行例 6-7，尝试 SQL 注入攻击，不会再出现图 6-3 和图 6-4 的严重后果。

6.8 存储过程

存储过程是存储在数据库中的一段已命名的 SQL 代码。利用存储过程，可以完成数据查询、插入、更新和删除等操作。存储过程可以接收输入参数，也可以将计算结果用参数输出。本节利用存储过程，实现"插入数据逻辑""查询数据逻辑"和"更新数据逻辑"，并给出简单的用户输入界面。

6.8.1 建立 SQL Server 的存储过程

进入 SQL Server 管理平台，右击 students 数据库，在弹出的快捷菜单中选择"可编程性"→"存储过程"→"新建存储过程"命令。存储过程建立后，需要"运行"和"刷新"才能在"存储过程"下看到建立的存储过程。

❶ 建立插入数据的存储过程 InsertStudent：

```
CREATE PROCEDURE InsertStudent
@student_no nchar(8),
@student_name nchar(15),
@email nchar(20),
@identification nchar(18),
@recordID int output
AS
if SUBSTRING(@identification,17, 1) % 2 = 0
   insert into student(student_no,student_name,gender,email,identification)
   values(@student_no,@student_name,'女',@email,@identification);
else
   insert into student(student_no,student_name,gender,email,identification)
```

```
values(@student_no,@student_name,'男',@email,@identification);
set @recordID = @@IDENTITY
```

说明：

① 存储过程使用 SUBSTRING 函数，根据身份证号判断性别，将值保存在 gender 字段中。

② 输出参数@recordID 返回插入新记录的 ID，该值在执行完 insert 语句后通过@@IDENTITY 函数得到。如果不使用存储过程，从刚插入的记录中获取自动生成的标识会非常麻烦。output 表示@recordID 是输出参数。

❷ 建立更新数据的存储过程 UpdateStudent：

```
CREATE PROCEDURE [dbo].[UpdateStudent]
@student_no   nchar(8),
@student_name nchar(15),
@email nchar(20),
@identification   nchar(18),
@id int
AS
if SUBSTRING(@identification,17, 1) % 2 = 0
   update student set student_no = @student_no, student_name = @student_name, gender = '女',
email = @email, identification = @identification where id = @id
else
   update student set student_no = @student_no, student_name = @student_name, gender = '男',
email = @email, identification = @identification where id = @id
```

❸ 建立删除数据的存储过程 DeleteStudent：

```
CREATE PROCEDURE [dbo].[DeleteStudent]
@id int
AS
    delete from student where id = @id
```

❹ 建立根据姓名和学号查询的存储过程 GetStudentByName 和 GetStudentByNo：

```
CREATE PROCEDURE GetStudentByName
@student_name char(15)
AS
select * from student where student_name = @student_name


CREATE PROCEDURE GetStudentByNo
@student_no char(8),
@student_count int output
AS
select @student_count = COUNT(*) from student where student_no = @student_no
```

说明：

① 存储过程 GetStudentByNo 根据输入的@student_no，判断该 student_no 的记录数。

② 输出参数@student_count 返回 student 中指定 student_no 的记录数。

6.8.2　在 DBbase 类中建立查询数据的逻辑

```
public static List<studentData> getStudentsByNameB(string student_name)
{
    //使用存储过程,根据姓名从 student 中查询
    List<studentData> student = new List<studentData>();
    using (SqlCommand cmd = new SqlCommand())
    {
        cmd.CommandText = "GetStudentByName";
        cmd.CommandType = CommandType.StoredProcedure;
        cmd.Connection = conn;
        cmd.Parameters.AddWithValue("@student_name",student_name);
        SqlDataReader reader = cmd.ExecuteReader();
        while (reader.Read())
        {
            student.Add(new studentData
            {
                student_no = reader["student_no"].ToString(),
                student_name = reader["student_name"].ToString(),
                email = reader["email"].ToString(),
                identification = reader["identification"].ToString(),
            });
        }
        reader.Close();
        return student;
    }
}
public static int getStudentCount(string student_no)
{
    //使用存储过程,根据学号从 student 中查询表中的记录数
    using (SqlCommand cmd = new SqlCommand("GetStudentByNo", conn))
    {
        cmd.CommandType = CommandType.StoredProcedure;
        cmd.Parameters.Add(new SqlParameter("@student_no",SqlDbType.Char,8));
        cmd.Parameters["@student_no"].Direction = ParameterDirection.Input;
        cmd.Parameters["@student_no"].Value = student_no;
        //命令参数集中增加另一个参数@student_count,是存储过程的输出参数
        cmd.Parameters.Add(new SqlParameter("@student_count", SqlDbType.Int, 5));
        cmd.Parameters["@student_count"].Direction = ParameterDirection.Output;
        //int student_count = (int) cmd.ExecuteScalar();
        cmd.ExecuteScalar();
        int student_count = (int) cmd.Parameters["@student_count"].Value;
        return student_count;
    }
}
```

说明：

① 命令对象默认情况下接受的是 SQL，故使用存储过程，需要指定命令对象的 CommandText 值为"存储过程名"，CommandType 值为 CommandType.StoredProcedure；

否则会出现异常。

② 虽然不使用输出参数,通过 int student_count=(int) cmd.ExecuteScalar();也能够获得记录数,本示例只是为了说明存储过程输入/输出参数的使用,所以专门使用了输出参数。

③ new SqlParameter("@student_no", SqlDbType.Char, 8)3 个参数分别是参数名、参数类型、参数大小。

④ (int) cmd.Parameters["@student_count"].Value;得到存储过程的输出参数 @student_count 的值。

6.8.3 在 DBbase 类中建立插入数据的逻辑

```
public static void InsertStudent(string student_no, string student_name, string email, string
identification,out int recordID)
{
    //使用存储过程,完成数据插入,并返回新插入记录的 ID
    using (SqlCommand cmd = new SqlCommand("InsertStudent", conn))
    {
        cmd.CommandType = CommandType.StoredProcedure;
        //为存储过程传入输入参数 "@student_no"
        SqlParameter param = new SqlParameter();
        param.ParameterName = "@student_no";
        param.Value = student_no;
        param.Direction = ParameterDirection.Input;    //是输入参数时可省略
        param.SqlDbType = SqlDbType.Char;
        cmd.Parameters.Add(param);
        //为存储过程传入输入参数 "@student_name"
        cmd.Parameters.Add(new SqlParameter("@student_name", SqlDbType.Char, 15));
        cmd.Parameters["@student_name"].Value = student_name;
        //为存储过程传入输入参数 "@email"
        cmd.Parameters.AddWithValue("@email", @email);
        //为存储过程传入输入参数 "@identification"
        cmd.Parameters.AddWithValue("@identification", identification);
        //为存储过程传入输出参数@recordID
        param = new SqlParameter();
        param.ParameterName = "@recordID";
        param.SqlDbType = SqlDbType.Int;
        param.Size = 5;
        param.Direction = ParameterDirection.Output;    //指定是输出参数,不可省略
        cmd.Parameters.Add(param);
        cmd.ExecuteNonQuery();
        //得到存储过程输出的参数值@recordID
        recordID = (int) cmd.Parameters["@recordID"].Value;
    }
}
```

说明:

① InsertStudent 通过使用修饰符 out 定义了一个带有输出的方法,在退出该方法前必须为该参数 recordID 赋值。

② 代码给出了 3 种不同传入参数的写法。当传入参数是输入参数时,参数的 Direction

属性可以不用设置。传入参数是输入参数时,使用 AddWithValue 方法,代码相对简练,其第 1 个参数是参数名,第 2 个参数是参数值。当传入的参数是输出参数时,必须指定 Direction 为 ParameterDirection.Output,但 AddWithValue 方法不能直接指定 Direction,需要再通过:命令对象.Parameters["@参数名"].Direction = ParameterDirection.Output 指定参数方向为输出。

6.8.4 在 DBbase 类中建立更新数据的逻辑

```
//重载 UpdateStudent 方法
public static void UpdateStudent(int id, string student_no, string student_name, string email, string identification)
{
    using (SqlCommand cmd = new SqlCommand("UpdateStudent", conn))
    {
        cmd.CommandType = CommandType.StoredProcedure;
        cmd.Parameters.AddWithValue("@id", id);
        cmd.Parameters.AddWithValue("@student_no", student_no);
        cmd.Parameters.AddWithValue("@student_name", student_name);
        cmd.Parameters.AddWithValue("@email", email);
        cmd.Parameters.AddWithValue("@identification", identification);
        cmd.ExecuteNonQuery();
    }
}
```

6.8.5 数据输入界面

例 6-9 新建 6_9.aspx 页面,直接使用 asp.net 控件,设计图 6-6 所示的页面。为了便于布局,可使用表格,将文本框放置在表格的单元格中。

界面中文本框 TxtStudent_no、TxtStudent_name、TxtEmail、TxtId 分别用于输入"学号""姓名""email""身份证号"。命令按钮的 Click 事件中的代码如下:

图 6-6 输入界面

```
if (DBbase.ConnectionSqlServer() == true)
{
    //先判断该学号是否已经输入,避免输入的主键不唯一时程序出现错误
    int student_count = DBbase.getStudentCount(TxtStudent_no.Text.Trim());
    if (student_count == 0)        //没有输入过该学号的记录
    {
        string student_no = TxtStudent_no.Text.Trim();
        string student_name = TxtStudent_name.Text.Trim();
        string email = TxtEmail.Text.Trim();
        string identification = TxtId.Text.Trim();
        int recordID = 0;
        DBbase.InsertStudent(student_no, student_name, email, identification, out recordID);
        Response.Write("<script>alert('新记录保存成功!')</script>");
    }
    else
    {
```

```
        Response.Write("<script>alert('该学号已经存在,重新输入')</script>");
    }
}
else
{
    Response.Write("<script>alert('连接数据库发生错误')</script>");
}
DBbase.CloseSqlServer();
```

说明:

① student 表中 student_no 是主键,不允许重复插入。数据插入前先调用查询逻辑 getStudentCount,判断相同的 student_no 是否已经存在。

② 调用带有输出参数的 InsertStudent 方法时,需要使用 out 修饰符。变量 recordID 使用前要预声明。通过设置断点,查看 recordID 值,其返回的是新增加记录的标识,即 student 表中自动编号字段 id 的值。

6.8.6 显示数据页面

ObjectDataSource 控件在网页控件和查询、更新的数据访问组件间建立了一个声明性的链接,它可以与 GridView 等控件协同工作。

例 6-10 基于查询逻辑,应用 GridView 和 ObjectDataSource 显示出 student 中全部记录。

❶ 新建名为 6_10.aspx 的文件,在设计界面放置 GridView 和 ObjectDataSource 控件,将 ObjectDataSource 控件的 ID 属性设置为 StudentSource。

❷ 设置 ObjectDataSource 控件的 TypeName 属性为 aspExample.DbEntity.DBbase (aspExample.DbEntity 是前面建立查询逻辑的命名空间,DBbase 是查询逻辑的类名)。设置 ObjectDataSource 控件的 SelectMethod 为 GetAllStudents(DBbase 类中建立的查询逻辑)。

❸ 设置 GridView1 控件的 DataSourceID 为 StudentSource,GridView1 自动根据数据逻辑中对 student 表的定义,更新各列。在数据逻辑更改后,通过"刷新架构"更新 GridView。

❹ Page_Load 事件中写入代码:

```
DBbase.ConnectionSqlServer();
```

❺ 运行程序,结果如图 6-7 所示。

student_no	student_name	gender	email	identification	id
41321059	李孝诚	男	1xc@163.com	110106199802030013	1
41340136	马小玉	女	mxy@126.com	130304199705113028	2
41355062	王长林	男	wcl@163.com	110106199512035012	3
41361045	李将寿	男	1jc@126.com	14510619901203503X	4
41361258	刘登山	男	1ds@126.com	130030520001234235	14
41361260	张文丽	女	zhangwl@163.com	140305200106202348	15
41401007	鲁宇星	男	1yx@sohu.com	110108200012035012	5
41405002	王小月	女	wxy@263.com	130304199901115047	7
41405003	张晨露	女	zcl@263.com	110107199810038022	6
41405005	张伟达	男	zwd@263.net	120305199010270018	10
41405221	张康林	女	zkl@126.net	130305199805231248	11

图 6-7 应用 ObjectDataSource 和 GridView 显示的数据

可以看出，建立好查询业务逻辑后，基本上不用编写任何程序代码就实现了数据的显示，用户界面和数据访问的代码是分离的。使用这两个控件，还可以完成数据查询、删除、更新等操作。

6.9 事务

事务作为一组数据库操作的一个整体，它们要么全部完成，要么全部失败。事务对于确保表中数据的安全性、有效性和一致性非常重要。

银行两个账户(A 和 B)间完成转账业务是非常经典的事务示例。从 A 账户转出 1000 元到 B 账户，必须保证按照以下事务方式发生。

(1) 从 A 账户的余额中减去 1000 元。

(2) B 账户上的余额增加 1000 元。

上面两个操作必须作为一个整体，才不至于从 A 账户减去 1000 元但没有到 B 账户的发生。事务中多个操作全部成功，事务才会被提交；如果任何一个操作失败，事务就会回滚到全部操作的原始状态。

ASP. NET Web 应用程序提供了 3 种类型的事务，代价由小到大的顺序如下。

(1) 存储过程事务。事务直接在存储过程中完成，由于只需要往返一次数据库，故这种事务性能最佳，缺点是需要用数据库的 SQL 编写事务处理的逻辑。

(2) 客户端引发的事务。利用 ADO.NET 提供的事务对象，在 ASP.NET 的页面中编写事务控制的代码。本质上与存储过程中的事务一样，使用相同的命令，但代码是使用封装了这些细节的 ADO.NET。缺点是事务开始和提交时，需要往返数据库。

(3) COM+事务。由 COM+运行时处理的一类事务。COM+事务采用两步提交协议，并且总会带来额外的开销。使用它们需要创建一个独立的服务组件类。因为它们支持分布式事务，所以只有当事务需要跨越多个资源管理器时才使用 COM+事务，如一个 COM+事务可以跨越 SQL Server 及 Oracle 数据库中的交互。

6.9.1 存储过程事务

事务存放的最佳地点是存储过程中，因为事务活动在数据库中，事务时间短，数据库并发性好，事务的代价小。存储过程的代码随着 DBMS 的不同而不同。不过大多数 RDBMS(关系型数据库管理系统)支持 BEGIN TRANSACTION 语句。一旦开始事务，其后所有的语句都被认为是事务的一部分，可以使用 COMMIT 或者 ROLLBACK 语句结束事务。如果不使用，事务将自动回滚。

例 6-11 假设事务如下：在 student 表中输入一条记录，在 score 表中也输入一条记录。在 SQL Server 中编写这个事务的存储过程。

建立存储过程 TransTest，代码如下：

```
CREATE PROCEDURE TransTest
@student_no char(8),
```

```
    @student_name char(15),
    @email char(20),
    @identification char(18),
    @course_no char(8),
    @score decimal(5,1)
AS
begin try
    begin transaction
        insert into student(student_no,student_name,email,identification)values(@student_no,
@student_name,@email,@identification)
        insert into score(student_no,course_no,score)values(@student_no,@course_no,@score)
    commit
end try
begin catch
    if(@@TRANCOUNT > 0)
        rollback
    declare @errMsg nvarchar(3000),@errSeverity int
    select @errMsg = ERROR_MESSAGE(),@errSeverity = ERROR_SEVERITY()
    raiserror(@errMsg,@errSeverity,1)
end catch
```

说明：

① begin try/begin catch 类似于 C♯ 的 try/catch 结构，begin try 中程序出现错误立即转到 begin catch。代码检查 @@TRANCOUNT 以确定当前是否有事务在进行中。变量 @@TRANCOUNT 计算当前连接中进行的事务数，begin transaction 使 @@TRANCOUNT 加 1，而 rollback 和 commit 使之减 1。

② ERROR_MESSAGE()错误消息，ERROR_SEVERITY()错误信息的级别。raiserror 语句，用于抛出一个异常或错误，类似于 asp.net 中的 throw new Exception。ADO.NET 把该消息传递给 SqlException 对象，可通过 .NET 代码捕捉。

可以在 SQL Server 中直接运行此存储过程测试事务：

```
exec transTest '40012030','赵三','zs@sohu.com','230304200010302378','k1',88
```

由于设置了 score 表与 class 表的级联关系，course_no='k1' 的课程在 class 表中不存在，故执行 insert into score(student_no,course_no,score)values(@student_no,@course_no,@score)语句插入记录时会出现错误，虽然存储过程先执行的是：insert into student(student_no,student_name,email,identification)values(@student_no,@student_name,@email,@identification)，但 student 中并没有插入新记录。如果运行：

```
exec transTest '40012030','赵三','zs@sohu.com','230304200010302378','k01',88
```

由于插入的记录没有违反级联规则，两个表中都正确地插入了新记录。

6.9.2 ADO.NET 事务

大多数 ADO.NET 数据提供程序对事务的支持。事务从调用 Connection 对象的 BeginTransaction 方法开始。

例 6-12 用 ASP.NET 编写事务代码，实现例 6-11。

❶ 建立存储过程 InsertScore 和插入 score 表数据的逻辑。

存储过程 InsertScore 代码：

```sql
CREATE PROCEDURE [dbo].[InsertScore]
@student_no nchar(8),
@course_no nchar(8),
@score decimal(5,1)
AS
insert into score(student_no,course_no,score) values(@student_no,@course_no,@score)
```

为 score 表插入数据的逻辑：

```csharp
public static void InsertScore(string student_no, string course_no, decimal score)
{
    //使用存储过程,完成数据插入,并返回新插入记录的 ID
    using (SqlCommand cmd = new SqlCommand("InsertScore", conn))
    {
        cmd.CommandType = CommandType.StoredProcedure;
        cmd.Parameters.AddWithValue("@student_no", student_no);
        cmd.Parameters.AddWithValue("@course_no", course_no);
        cmd.Parameters.AddWithValue("@score", score);
        cmd.ExecuteNonQuery();
    }
}
```

❷ 在 DBbase 类中建立事务处理的逻辑。

```csharp
public static void TransTest(string student_no, string student_name, string email, string identification, string course_no, decimal score, out int recordID)
{
    //为 student 表插入记录
    SqlCommand cmdA = new SqlCommand("InsertStudent", conn);
    cmdA.CommandType = CommandType.StoredProcedure;
    cmdA.Parameters.AddWithValue("@student_no", student_no);
    cmdA.Parameters.AddWithValue("@student_name", student_name);
    cmdA.Parameters.AddWithValue("@email", email);
    cmdA.Parameters.AddWithValue("@identification", identification);
    cmdA.Parameters.Add(new SqlParameter("@recordID", SqlDbType.Int));
    cmdA.Parameters["@recordID"].Direction = ParameterDirection.Output;
    //为 score 表插入记录
    SqlCommand cmdB = new SqlCommand("InsertScore", conn);
    cmdB.CommandType = CommandType.StoredProcedure;
    cmdB.Parameters.AddWithValue("@student_no", student_no);
    cmdB.Parameters.AddWithValue("@course_no", course_no);
    cmdB.Parameters.AddWithValue("@score", score);
    SqlTransaction trans = null;
    try
    {
        trans = conn.BeginTransaction();
        cmdA.Transaction = trans;
```

```
            cmdB.Transaction = trans;
            cmdA.ExecuteNonQuery();
            cmdB.ExecuteNonQuery();
            recordID = (int)cmdA.Parameters["@recordID"].Value;
            //提交事务
            trans.Commit();
        }
        catch (SqlException errMsg)
        {
            trans.Rollback();
            recordID = 0;
        }
        finally
        {
            conn.Close();
        }
    }
```

说明：仅仅创建和提交事务是不够的,还需要设置 Command.Transaction 对象,从而将 Command 对象列入事务中。如果在事务进行中执行一个不在当前事务中的命令,会产生错误。

❸ 在页面 Load 事件中写入以下代码：

```
int recordID;
if (DBbase.ConnectionSqlServer() == true)
{
    DBbase.TransTest("30012036", "张雨", "zy@sohu.com", "230304200210052346", "k01", 88, out recordID);
}
```

程序运行后,在 student 表和 score 表中同时正确地插入了记录。尝试将上面程序中对应的语句改为：

```
DBbase.TransTest("30012040", "张雨", "zy@sohu.com", "230304200210052346", "k88", 88, out recordID);
```

student 表中 student_no="30012040"的记录不存在,在不使用事务的情况下,可以将该条记录插入到 student 表中。由于设置了 course 表和 score 表的级联规则, course 表中不存在 course_no="k88"的记录, score 表中插入该记录时违反了级联规则。将两个插入动作放在一个事务中,数据提交失败,事务回滚,两个表中都没有输入数据。

使用事务时的注意事项如下。

① 事务要尽量短。

② 不要在事务中使用 select 查询返回数据,就在事务前返回数据,以减少事务锁定记录的数目。

③ 如果事务中需要查询记录,只查询确实需要的记录。

④ 尽量使用存储过程的事务,而不是使用 ADO.NET 事务。

⑤ 尽量避免大批量数据的同时更新。

⑥ 尽量避免使用具有多个独立批处理任务的事务,把各个批处理任务作为单个事务。

习题与思考

(1) 使用 DataReader 列出 student 表中所有字段名。

(2) 利用 6.5 节程序的漏洞,在图 6-4 所示的文本框中输入什么内容,攻击者就可以轻易地将 student 库从服务器上删除?

(3) 将 6.5.4 小节和 6.5.5 小节中的"插入数据逻辑"和"更新数据逻辑"都改为参数命令对象。

(4) 新建存储过程 DeleteStudentByNo,根据 student_no 删除记录;为存储过程 DeleteStudentByNo 建立删除数据的逻辑,并设计简单的删除界面,根据输入的学号删除 student 表中记录。

(5) 编写一段 C#代码,调用例 6-11 中的存储过程 TransTest 测试事务。

第 7 章

非连接的数据访问对象和工厂模型

(1) DataAdapter、DataSet、DataTable、DataView 的属性和方法。
(2) DataTable 数据的读取和序列化。
(3) DataSet 中表间关系的建立。
(4) ADO.NET 的工厂模型。

第 6 章通过 ADO.NET 访问数据库,使用的主要对象是 Connection、Command 和 DataReader,这些对象是基于连接的对象。本章将介绍基于非连接的数据访问对象 DataAdapter、DataSet、DataTable 和 DataView。

ADO.NET 连接数据库后,用数据库中的信息填充 DataSet,然后断开与数据库的连接。如果修改了 DataSet 中的信息,原来数据库中对应的信息并不会随之改变,当然如果需要修改原数据库中的信息,可以重新连接数据库,通过批操作把 DataSet 中的数据修改到数据库中。

以下情景更适合于使用非连接访问模型和 DataSet。

(1) 从大量的数据中前后浏览数据。由于在线连接的 DataReader 只支持向前移动,而 DataSet 由于支持分页和前后移动,浏览更加方便。

(2) 需要在不同表间导航。DataSet 可以保持其所有表及这些表的关系,故使用 DataSet 可以建立一对多的页面而不需要多次到数据库中执行查询。

(3) 需要将数据同时绑定到页面的多个控件上,需要对数据自定义排序和筛选。DataReader 是只读向前的,故不能同时绑定在多个控件上,也不能动态排序和筛选,而 DataSet 却可以。

(4) 需要以 XML 方式操作数据时。

(5) 需要通过 Web 方式批量更新数据时。客户端下载数据到 DataSet,修改了多个数据,然后提交服务器,更新原数据库。这种方式的不足之处是要考虑冲突的发生。

(6) 需要将数据以某种格式(如 XML)保存到磁盘的情况下。

（7）需要包在组件间方便地传递时。

7.1 DataSet 类

DataSet 是非连接数据访问的核心，是关系数据在内存中的一种表现形式。DataSet 类的内部包含了 3 个强类型（为所有变量指定数据类型）的集合，即 DataTableConnection、DataRelationConnection 和 PropertyCollection，如图 7-1 所示。

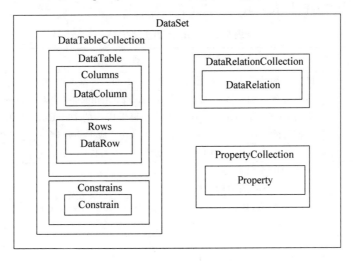

图 7-1 DataSet 对象模型

DataSet 中可以存放若干数据表对象，而每个数据表中的数据是由 DataAdapter 的 Fill() 方法填充。每个数据表还有数据行（DataRow），每个数据行又有多个数据列（DataColumn）。DataSet 中多个表间的关系，可以通过编程，利用 DataRelationCollection 表示各表间的这种关系，这种关系类似于关系数据库中的主键列与外键列间的连接，DataRelation 标识 DataSet 中两个表的匹配列。

DataSet 的 ExtendedProperties 属性提供了对 PropertyCollection 对象的访问，通过 ExtendedProperties 属性可以将一些新的属性加入到 DataSet 中。

7.2 DataSet 类的主要属性和方法

7.2.1 DataSet 的主要属性

DataSet 的主要属性见表 7-1。

表 7-1 DataSet 的主要属性

属　性	作　用
CaseSensitive	DataTable 中的字符串比较时是否区分大小写，默认为 false
DataSetName	表示 DataSet 的一个别名
EnforceConstrains	执行更新操作时是否遵循约束规则

续表

属性	作用
HasError	提示在 DataSet 中所有 DataTable 的数据行是否存在问题
RemotingFormat	定义 DataSet 在序列化时采用的方式,默认是 XML

7.2.2 DataSet 的主要方法

DataSet 的主要方法见表 7-2。

表 7-2 DataSet 的主要方法

方法	作用
AcceptChanges()	提交自加载此 DataSet 或上次调用 AcceptChanges 以来对其进行的所有更改
RejectChanges()	回滚自该表加载以来或上次调用 AcceptChanges 以来对该表进行的所有更改。调用 RejectChanges 时,任何仍处于编辑模式的 DataRow 对象将取消其编辑,新行被移除,DataRowState 设置为 Modified 或 Deleted 的行返回到其初始状态
Clear()	移除所有 DataTable 中的行,清除所有 DataTable 中的记录
Clone()	复制 DataSet 的结构
Copy()	复制 DataSet 的结构和数据
HasChanges()	获取一个值,该值指示 DataSet 是否有 DataRowState 被筛选的更改,包括新增行、已删除的行或已修改的行。获取的值包括 DataRowState. Added、DataRowState. Detached、DataRowState. Unchanged、DataRowState. Deleted、DataRowState. Modified
GetChanges()	获取 DataSet 的副本,该副本包含自加载以来或自上次调用 AcceptChanges 以来对该数据集进行的所有更改。在调用 GetChanges()方法之前,可先检查 DataSet 的 HasChanges
ReadXML()	读取 XML
WriteXML()	将 DataSet 的内容写入 XML 文件

例 7-1 练习 DataSet、DataTable 中记录的增加以及 HasChanges()、GetChanges()、AcceptChanges()、RejectChanges()等的使用。

新建 7_1.aspx Web 窗体文件,新增加引入的命名空间:

```
using System.Data.SqlClient;
using System.Web.Configuration;
```

```
1   public partial class _7_1 : System.Web.UI.Page
2   {
3       string msg = "";
4       DataSet tmpDataSet = new DataSet();
5       protected void Page_Load(object sender, EventArgs e)
6       {
7           string cnnString = WebConfigurationManager.ConnectionStrings["SQLServerString"].ConnectionString;
8           SqlConnection conn = new SqlConnection(cnnString);
9           DataSet ds = new DataSet();
10          SqlDataAdapter adapter = new SqlDataAdapter("select * from student", conn);
```

```
11      adapter.Fill(ds, "student");
12      DataTable dt = ds.Tables["student"];
13      DataRow newRow = dt.newRow();
14      newRow["student_no"] = "1";
15      newRow["student_name"] = "赵1";
16      dt.Rows.Add(newRow);
17      ds.RejectChanges();
18      newRow = dt.newRow();
19      newRow["student_no"] = "2";
20      newRow["student_name"] = "赵2";
21      dt.Rows.Add(newRow);
22      ds.AcceptChanges();
        //只有下面两条记录才能更新到数据库中
23      newRow["student_no"] = "3";
24      newRow["student_name"] = "赵3";
25      dt.Rows.Add(newRow);
26      newRow["student_no"] = "4";
27      newRow["student_name"] = "赵4";
28      dt.Rows.Add(newRow);
29      if (!ds.HasChanges()) return;
30      tmpDataSet = ds.GetChanges(DataRowState.Added);
31      if (tmpDataSet == null) return;
32        dt.RowChanged += new DataRowChangeEventHandler(Row_Changed);
33        SqlCommandBuilder cmdBuilder = new SqlCommandBuilder(adapter);
34        if (!tmpDataSet.HasErrors)
35        {
36            msg = "数据插入成功!";
37            adapter.Update(tmpDataSet.Tables[0]);
38        }
39        else
40        {
41            msg = "数据插入发生错误!";
42        }
43      }
44      private void Row_Changed(object sender, DataRowChangeEventArgs e)
45      {
46          Response.Write("<script>alert('" + msg + "')</script>");
47      }
48  }
```

说明：

① 第 9 行定义一个 DataSet 对象 ds，第 10 行定义 SqlDataAdapter 对象 adapter。

② 第 11 行使用 DataAdapter 的 Fill 方法，为 ds 中的 DataTable 充填数据，充填后的表名为 student。

③ 第 12 行将 ds 中的 student 赋值给 DataTable 类型的对象 dt，第 13 行向 dt 中插入一条空的数据行 newRow。

④ 第 14~16 行为数据行 newRow 的列 student_no 和 student_name 的赋值；第 18~21 行、第 23~28 行也都是为数据行 newRow 增加新的记录。由于第 17 行 ds.RejectChanges() 和第 22 行 AcceptChanges()，故只有第 23~28 行增加的两条记录能够通过后面的命令更

新到数据库中。

⑤ 第 29 行判断 ds 是否有增加、更新、删除记录等操作,第 30 行将新增加记录的 ds 复制到另一个 DataSet——tmpDataSet 中。

⑥ 第 32 行为 dt 增加一个事件过程 RowChanged,该事件在 dt 增加记录完成后触发;第 44~47 行定义了该事件过程。

⑦ 使用数据适配器对数据库更新记录时,需要使用 CommandBuilder 对象,以便让系统自动生成对应维护的 Insert、Update、Delete 操作的命令对象。

⑧ 第 37 行调用 DataAdapter 对象的 Update 方法以执行 CommandBuilder 对象生成的 3 个命令对象。

7.3 DataTable 类

7.3.1 DataTable 的使用

在使用 DataAdapter 和 DataSet 时,往往需要涉及 DataTable,而 DataTable 是由 DataColumn 和 DataRow 构成的,故需要掌握这些对象的使用方法。

如果以编程方式创建 DataTable,必须先创建 DataColumn 对象,将其加入到 DataColumnCollection 中定义其架构,类似于定义表结构。

如果要向 DataTable 添加行,必须先使用 NewRow()方法,返回新的 DataRow 对象。NewRow()方法返回具有 DataTable 架构的行,DataTable 可存储的最大行数是 16777216。

DataTable 包含了由表的 DataColumns 属性引用的 DataColumn 对象的集合,列的集合与任何约束一起构成表的架构。

例 7-2 定义一个 DataTable,包含有 Sno、Sname、Sex、Birthday 共 4 列,增加一条记录,并且用 GriewView1 控件显示出该记录。

```
//定义一个 DataTable 的对象 dt
DataTable dt = new DataTable("student");
//定义 DataColumn 的实例对象 column,其名称为 sno,类型是字符串,typeof 操作符返回数据类型
DataColumn column = new DataColumn("sno", typeof(string));
//该列不能为空
column.AllowDBNull = false;
//该列值唯一
column.Unique = true;
//将定义的列加入到 DataTable 对象 dt 中
dt.Columns.Add(column);
column = new DataColumn("sname",typeof(string));
dt.Columns.Add(column);
column = new DataColumn("sex", typeof(string));
dt.Columns.Add(column);
//定义 DataColumn 的实例对象 column,名称为 birthday,类型是日期型
column = new DataColumn("birthday", typeof(DateTime));
dt.Columns.Add(column);
//以下内容为 DataTable 增加一条记录
DataRow newRow = dt.NewRow();
```

```
newRow["sno"] = "A2010221";
newRow["sname"] = "刘东明";
newRow["sex"] = "男";
newRow["birthday"] = "2010-7-5";
dt.Rows.Add(newRow);
//将 dt 显示在 GridView1 中
GridView1.DataSource = dt;
GridView1.DataBind();
```

DataTable 的主要属性见表 7-3。

表 7-3　DataTable 的主要属性

属　　性	作　　用
Rows	DataTable 中所有行的集合
Columns	DataTable 中所有列的集合
CaseSensitive	表中字符串比较时,是否区分大小写,默认为 false
ChildRealations	获取此 DataTable 的子关系的集合(如果有)
Constrains	获取该表维护的约束的集合
PrimaryKey	获取或设置数据表的主键的列的数组

7.3.2　获取 DataTable 中的数据

1. 使用 Rows 和 Columns 属性获取数据

例 7-3　列出 student 表中的记录,输出表的列名和记录。

❶ 新建文件 7_3.aspx,在 7_3.aspx.cs 中引入命名空间:

```
using System.Data;
using System.Web.Configuration;
using System.Data.SqlClient;
using System.Data.Common;
```

❷ 在 Page_Load 事件中写入以下代码:

```
string cnnString = WebConfigurationManager.ConnectionStrings["SQLServerString"].ConnectionString;
SqlConnection conn = new SqlConnection(cnnString);
string strSQL = "select * from student";
SqlDataAdapter adapter = new SqlDataAdapter(strSQL, conn);
DataSet ds = new DataSet();
adapter.Fill(ds, "student");
DataTable dt = ds.Tables["student"];
//列出字段名
Response.Write("<table style = 'text-align:center'><tr>");
for (int i = 0; i < dt.Columns.Count; i++)
{
    Response.Write("<td>" + dt.Columns[i].ColumnName + "</td>");
}
Response.Write("</tr>");
//列出字段值
for (int i = 0; i < dt.Rows.Count; i++)
```

```
        {
            Response.Write("<tr>");
            for (int j = 0; j < dt.Columns.Count; j++)
            {
                Response.Write("<td>" + dt.Rows[i][j].ToString() + "</td>");
            }
            Response.Write("</tr>");
        }
        Response.Write("</table>");
```

说明：

① 示例采用了<table><tr><td></td></tr></table>控制页面内容输出的格式。

② 通过使用 dt.Columns[i].ColumnName 索引器输出列名(字段名)。

③ 使用 t.Rows[i][j]输出 i 行 j 行的字段值。

程序运行后，输出结果如图 7-2 所示。

student_no	student_name	gender	email	identification	id
30012035	楚云飞	男	zsf@sohu.com	2303042006112011537	1
40012030	赵三	男	zs@sohu.com	2303042000103102378	3
41321059	李孝诚	男	lxc@163.com	110106199802030013	4
41340136	马小玉	女	mxy@126.com	130304199705113028	5
41355062	王长林	男	wcl@163.com	110106199512035012	6
41361045	李将寿	男	ljc@126.com	14510619901203503X	7
41361258	刘登山	男	lds@126.com	130030520001234235	8
41361260	张文丽	女	zhangwl@163.com	140305200106202348	9
41401007	鲁宇星	男	lyx@sohu.com	110108200012035012	10
41405002	王小月	女	wxy@263.com	130304199901115047	11
41405003	张晨露	女	zcl@263.com	110107199810038022	12
41405005	张伟达	男	zwd@263.net	120305199010270018	13
41405221	张康林	女	zkl@126.net	130305199805231248	14
41405238	张雨	女	zy@sohu.com	230304200210052346	2
A9401843	赵飞强				35

图 7-2　程序运行结果

如果要改变输出的标题行，如使标题行变为中文，方法是将示例中 string strSQL = "select * from student";改为 string strSQL = "select student_no as 学号,student_name as 姓名 from student"，也可以通过映射数据库名的方法加以实现，使用此功能，需要引入命名空间 using System.Data.Common;，将以下代码放在 DataSet ds = new DataSet()语句前面：

```
DataTableMapping studentMap = adapter.TableMappings.Add("student", "student");
studentMap.ColumnMappings.Add("student_no", "学号");
studentMap.ColumnMappings.Add("student_name", "姓名");
studentMap.ColumnMappings.Add("email", "电子邮件");
studentMap.ColumnMappings.Add("gender", "性别");
studentMap.ColumnMappings.Add("identification", "身份证号");
studentMap.ColumnMappings.Add("id", "自动编号");
```

再次运行程序，结果见图 7-3。

本示例没有直接将 DataTable 绑定在控件如 GridView 上，而是采用 DataTable 的

学号	姓名	性别	电子邮件	身份证号	自动编号
30012035	楚云飞	男	zsf@sohu.com	230304200611201537	1
40012030	赵三	男	zs@sohu.com	230304200010302378	3
41321059	李孝诚	男	lxc@163.com	110106198802030013	4
41340136	马小玉	女	mxy@126.com	130304199705113028	5
41355062	王长林	男	wcl@163.com	110106199512035012	6
41361045	李将寿	男	ljc@126.com	14510619901203503X	7
41361258	刘登山	男	lds@126.com	130030520001234235	8
41361260	张文丽	女	zhangwl@163.com	140305200106202348	9
41401007	鲁宇星	男	lyx@sohu.com	110108200012035012	10
41405002	王小月	女	wxy@263.com	130304199901115047	11
41405003	张晨露	女	zcl@263.com	110107199810038022	12
41405005	张伟达	男	zwd@263.net	120305199010270018	13
41405221	张康林	女	zkl@126.net	130305199805231248	14
41405238	张雨	女	zy@sohu.com	230304200210052346	2
A9401843	赵飞强				35

图 7-3　利用映射数据库改变字段显示的名称

Columns 和 Rows 属性，利用其提供的索引器，输出字段名和字段值。利用映射数据库可以将输出的字段改名，使其更加容易理解。

2. 使用 DataTableReader 对象读取 DataTable 的数据

DataTable 提供了一个 CreateDataReader() 的方法，该方法允许使用像 DataReader 一样的方式，从内存中的 DataTable 中读取记录，而不是从实际的数据库中读取记录：

```
string cnnString = WebConfigurationManager.ConnectionStrings["SQLServerString"].ConnectionString;
SqlConnection conn = new SqlConnection(cnnString);
string strSQL = "select * from student";
SqlDataAdapter adapter = new SqlDataAdapter(strSQL, conn);
DataSet ds = new DataSet();
adapter.Fill(ds, "student");
DataTable dt = ds.Tables["student"];
DataTableReader dtReader = dt.CreateDataReader();
while (dtReader.Read())
{
    for (int i = 0; i < dtReader.FieldCount; i++)
    {
        Response.Write(dtReader.GetValue(i).ToString().Trim());
    }
    Response.Write("<br>");
}
```

7.3.3　DataTable 中删除和更新记录

例 7-4　结合 GridView，可视化地删除和更新 DataTable 中的记录。
示例需要引入以下命名空间：

```
using System.Data;
using System.Web.Configuration;
using System.Data.SqlClient;
```

❶ 新建 Web 窗体 7_4.aspx,在上面放置控件 GridView1、TextBox1 和 Button1 及 Label1,将 Label1 的 Text 设置为"输入或选择要删除的学号"。

❷ 定义一个 Web 窗体的公共变量:

```
static string cnnString = WebConfigurationManager.ConnectionStrings["SQLServerString"].ConnectionString;
static SqlConnection conn = new SqlConnection(cnnString);
static SqlDataAdapter adapter = new SqlDataAdapter();
static DataTable dt = new DataTable();
static DataSet ds = new DataSet();
```

❸ 在 Page_Load 事件中写入以下代码:

```
if (!Page.IsPostBack)
{
    string strSQL = "select * from student";
    adapter = new SqlDataAdapter(strSQL, conn);
    adapter.SelectCommand = new SqlCommand(strSQL, conn);
    ds = new DataSet();
    adapter.Fill(ds, "student");
    dt = ds.Tables["student"];
    GridView1.DataSource = dt;
    GridView1.DataBind();
}
```

❹ 在 GridView1_RowDataBound() 中的代码如下:

```
protected void GridView1_RowDataBound(object sender, GridViewRowEventArgs e)
{
    if (e.Row.RowType == DataControlRowType.DataRow)
    {
        e.Row.Attributes.Add("style", "cursor:pointer");
        e.Row.Attributes.Add("onclick", "document.getElementById('TextBox1').value = " + e.Row.Cells[0].Text + "");
    }
}
```

说明:GridView1_RowDataBound()事件在给 GridView 控件绑定完数据后触发,e.Row.RowType == DataControlRowType.DataRow 判断是否是数据行(还可能有标题行、脚注行等)。e.Row.Attributes.Add()在行里面增加属性 e.Row.Attributes.Add("style","cursor:pointer");改变鼠标形状,语句 e.Row.Attributes.Add("onclick", "document.getElementById('TextBox1').value=" + e.Row.Cells[0].Text+"");为行增加单击事件:当单击时使 TextBox1 的值等于 GridView1 当前行第 1 值,即 student_no 的值。

❺ 选择"设计"视图中的命令按钮,在命令按钮的 OnClientClick 属性中输入:

```
return confirm('真的要删除吗?')
```

说明:OnClientClick 表示在客户端 OnClick 上执行的客户端脚本,如果脚本返回是 false,将停止向服务器提交数据。confirm()是 JavaScript 的 window 对象的一个方法,出现对话框后,单击"确定"按钮,返回 true,单击"取消"按钮返回 false。

❻ 在 Button1 的 Click 事件中写入以下代码：

```
DataRow[] rowToDelete = dt.Select(string.Format("student_no = '{0}'",TextBox1.Text));
rowToDelete[0].Delete();
GridView1.DataSource = dt;
GridView1.DataBind();
```

说明：DataTable 的 Select()方法可以指定一个搜索条件，类似于 SQL 语法，返回匹配搜索条件的 DataRow 对象的数组。rowToDelete[0].Delete();输入 0 是因为 student_no 是唯一的，故搜索后的结果只有一个。删除数据后，需要重新将 dt 绑定到 GridView1。

程序运行结果见图 7-4。

student_no	student_name	gender	email	identification	id
30012035	楚云飞	男	zsf@sohu.com	230304200611201537	1
40012030	赵三	男	zs@sohu.com	230304200010302378	3
41321059	李孝诚	男	lxc@163.com	110106199802030013	4
41340136	马小玉	女	mxy@126.com	130304199705113028	5
41355062	王长林	男	wcl@163.com	110106199512035012	6
41361045	李将寿	男	ljc@126.com	14510619901203503X	7
41361258	刘登山	男	lds@126.com	130030520001234235	8
41361260	张文丽	女	zhangwl@163.com	140305200106202348	9
41401007	鲁宇星	男	lyx@sohu.com	110108200012035012	10
41405002	王小月	女	wxy@263.com	130304199901115047	11
41405003	张晨露	女	zcl@263.com	110107199810038022	12
41405005	张伟达	男	zwd@263.net	120305199010270018	13
41405221	张康林	女	zkl@126.net	130305199805231248	14
41405238	张雨	女	zy@sohu.com	230304200210052346	2
455	x444	男	abc@126.com	123456789012	36

输入或选择要删除的学号：455 删除

图 7-4 示例运行结果

图 7-4 中将光标移动到某一条记录上单击，该记录的 student_no 值将出现在文本框中，单击"删除"按钮，出现对话框，单击"确认"按钮后，选中记录将从 dt 中移除；单击"取消"按钮后放弃删除。

按照这种方法，删除某条记录后，检查数据库中 student 表会发现，记录并没有从数据库中删除，而只是从当前内存中的 dt 中删除了。如果要采用这种方法，从数据库中删除选定的记录，需要在命令按钮中增加以下两条语句：

```
SqlCommandBuilder scb = new SqlCommandBuilder(adapter);
adapter.Update(ds,"student");
```

说明：CommandBuilder 可以为 DataAdapter 自动构建更新数据库的 InsertCommand、UpdateCommand、DeleteCommand 命令，但有以下几个条件必须满足。

① 更新的表只能是单一的表。

② 预先要为 DataAdapter 指定 SelectCommand 命令，根据 SelectCommand，CommandBuilder 才能自动生成其他更新数据库的命令。

③ 更新的数据表中必须定义了主键。

7.4 DataAdapter 类

要将数据库中的数据提取到 DataSet，需要借助 DataAdapter，DataAdapter 是联系物理数据库和 DataSet 的桥梁，它含有查询和更新数据源的全部命令。

7.4.1 使用 DataAdapter 填充 DataSet

例 7-5 使用 DataSet、DataAdapter,将 student 表中数据显示在 Gridview 控件中。

❶ 新建 Web 窗体 7_5.aspx,在 7_5.aspx.cs 中引入命名空间:

```
using System.Web.Configuration;
using System.Data.SqlClient;
using System.Data;
```

❷ 在窗体上放置控件 GridView1,在 Page_Load 事件中写入以下代码:

```
string cnnString = WebConfigurationManager.ConnectionStrings["SQLServerString"].ConnectionString;
SqlConnection conn = new SqlConnection(cnnString);
string strSQL = "select * from student";
SqlDataAdapter adapter = new SqlDataAdapter(strSQL, conn);
DataSet ds = new DataSet();
adapter.Fill(ds);
GridView1.DataSource = ds.Tables[0];
GridView1.DataBind();
```

说明:

① 定义 DataAdapter 时 SqlDataAdapter 的第 1 个参数是传入的 SQL 语句,第 2 个参数是连接的对象名;使用 DataAdapter 时不用显式地连接数据库,其能够自动连接数据库,并且在数据充填完成后能够自动断开与数据库的连接。如果是手动打开连接则需要手动关闭连接。

② adapter.Fill(ds)充填数据到 DataSet,充填后的表名通过 ds.Tables[0]、ds.Tables[1]的方式指定,其中 0,1,……,n 是充填后表的序列编号,也可以显式地为充填得到的表命名,如 adapter.Fill(ds, "student")将充填后的表显式地命名为 student。

③ 上述代码并没有打开 Connection 对象,而是当调用 Fill()方法时由 DataAdapter 自动打开和关闭 Connection。

④ 由于 DataAdapter 填充 DataSet 采用的是 adapter.Fill(ds),没有指定表名,故需要 GridView1.DataSource = ds.Tables[0];,如果 adapter.Fill(ds, "aaaa")指定了填充后的表名,则可用代码 GridView1.DataSource=ds.Tables[0];或者 GridView1.DataSource=ds.Tables["aaaa"];。

DataAdapter 内部封装了 4 个命令对象,即 SelectCommand、InsertCommand、DeleteCommand、UpdateCommand,借助这 4 个对象,DataAdapter 能够完成各种对数据库的操作。

例 7-5 中的代码也可以更改为:

```
string cnnString = WebConfigurationManager.ConnectionStrings["SQLServerString"].ConnectionString;
SqlConnection conn = new SqlConnection(cnnString);
string strSQL = "select * from student";
SqlDataAdapter adapter = new SqlDataAdapter();
adapter.SelectCommand = new SqlCommand(strSQL,conn);
DataSet ds = new DataSet();
adapter.Fill(ds);
```

```
GridView1.DataSource = ds.Tables[0];
GridView1.DataBind();
```

7.4.2 将 DataSet/DataTable 对象序列化为 XML

DataSet 和 DataTable 都提供了 WriteXML()和 ReadXML()方法，WriteXML()将 DataSet 或 DataTable 中的内容持久化成 XML 形式的本地文件，ReadXML()允许从 XML 文档加载数据到 DataSet 或 DataTable。DataSet 和 DataTable 都支持 WriteXmlSchema()和 ReadXMLSchema()保存和加载一个*.XSD 文件。

```
public void SaveToXML(DataTable dt)
{
    dt.DataSet.DataSetName = "students";
    dt.WriteXml(Server.MapPath("student.xml"));
    dt.WriteXmlSchema(Server.MapPath("student.xsd"));
}
```

说明：

① 没有 dt.DataSet.DataSetName = "students";，生成的 XML 的根节点为<NewDataSet>。

② Server 是 ASP.NET 提供的一个内部对象，DataSet 和 DataTable 序列化 XML 文件时，需要提供 XML 文件的物理路径。使用 Server.MapPath 将不带路径的文件名转换为带物理路径的文件名。dt.WriteXML(Server.MapPath("student.xml"))生成的 student.xml 和当前网页在同一个文件夹中。

调用 SaveToXML()后，生成的 student.xml 内容如下：

```xml
<?xml version="1.0" standalone="yes"?>
<students>
  <student>
    <student_no>30012035</student_no>
    <student_name>楚云飞</student_name>
    <gender>男</gender>
    <email>zsf@sohu.com</email>
    <identification>230304200611201537</identification>
    <id>1</id>
  </student>
  <student>
    <student_no>40012030</student_no>
    <student_name>赵三</student_name>
    <gender>男</gender>
    <email>zs@sohu.com</email>
    <identification>230304200010302378</identification>
    <id>3</id>
  </student>
   ...
</students>
```

在 VS 中打开 student.xsd，将打开 IDE 的 XML 架构编辑器，如图 7-5 所示。

图 7-5 Visual Studio 的 XSD 编辑器

7.4.3 将 DataSet/DataTable 对象以二进制格式序列化

虽然 XML 可以跨平台传递，但由于其内部包含大量的描述性标记，无形中会增加网络的负担。而将 DataSet/DataTable 保存为二进制文件，则可以减轻网络负担。

要以二进制格式序列化 DataSet/DataTable，需要设置其属性 RemotingFormat 为 SerializationFormat.Binary。需要引入的命名空间如下：

```
using System.IO;
using System.Runtime.Serialization.Formatters.Binary;
public void BinarySerialization(DataSet ds)
{
    //设置二进制序列化标记
    ds.RemotingFormat = SerializationFormat.Binary;
    //以二进制格式保存 DataSet
    FileStream fs = new FileStream(Server.MapPath("students.bin"),FileMode.Create);
    BinaryFormatter bformat = new BinaryFormatter();
    bformat.Serialize(fs,ds);
    fs.Close();
    //将生成的 students.bin 反序列化为 DataSet
    fs = new FileStream(Server.MapPath("students.bin"), FileMode.Open);
    DataSet myDS = (DataSet)bformat.Deserialize(fs);
}
```

将生成的 students.bin 文件添加到"解决方案资源管理器"中，用 VS 打开该文件，显示如图 7-6 所示，可以看到 students.bin 是一个二进制文件。

如果二进制序列化的是 DataTable 对象，如 dt，必须先指定 DataSet 的 RemotingFormat，然后再指定 dt 的 RemotingFormat。从表 7-1 中可以看出 DataSet 默认的序列化方式是 XML，直接指定 DataTable 的序列化方式为二进制，会出现 DataTable 与 DataSet 指定的 RemotingFormat 不一致的错误提示。DataTable 序列化时，同时指定序列化方式的代码，指定代码如下：

```
dt.DataSet.RemotingFormat = SerializationFormat.Binary;
dt.RemotingFormat = SerializationFormat.Binary;
```

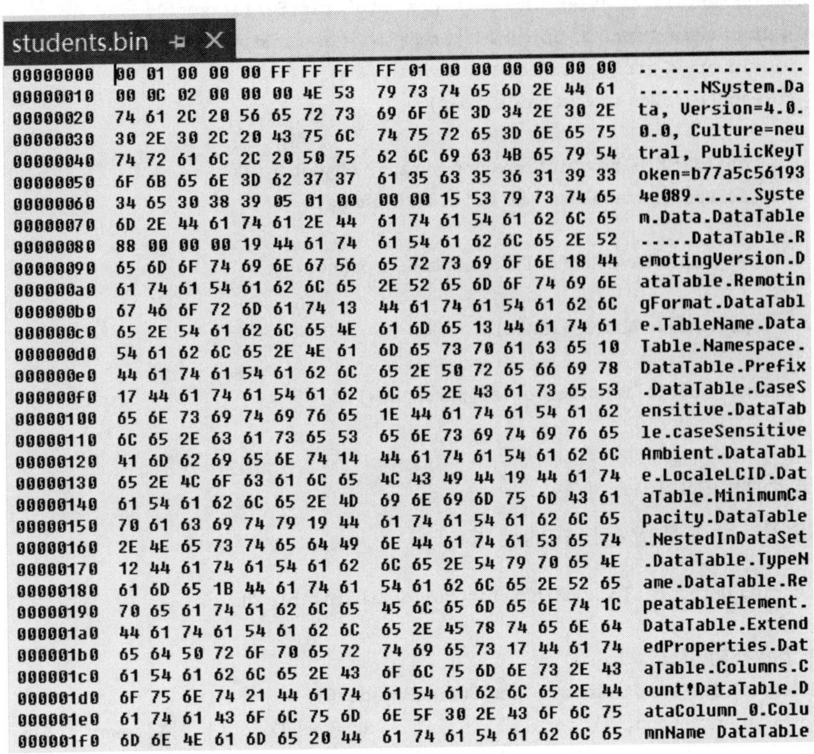

```
students.bin
00000000  00 01 00 00 00 FF FF FF  FF 01 00 00 00 00 00 00  ................
00000010  00 0C 02 00 00 00 4E 53  79 73 74 65 6D 2E 44 61  ......NSystem.Da
00000020  74 61 2C 20 56 65 72 73  69 6F 6E 3D 34 2E 30 2E  ta, Version=4.0.
00000030  30 2E 30 2C 20 43 75 6C  74 75 72 65 3D 6E 65 75  0.0, Culture=neu
00000040  74 72 61 6C 2C 20 50 75  62 6C 69 63 4B 65 79 54  tral, PublicKeyT
00000050  6F 6B 65 6E 3D 62 37 37  61 35 63 35 36 31 39 33  oken=b77a5c56193
00000060  34 65 30 38 39 05 01 00  00 00 15 53 79 73 74 65  4e089......Syste
00000070  6D 2E 44 61 74 61 2E 44  61 74 61 54 61 62 6C 65  m.Data.DataTable
00000080  88 00 00 00 19 44 61 74  61 54 61 62 6C 65 2E 52  .....DataTable.R
00000090  65 6D 6F 74 69 6E 67 56  65 72 73 69 6F 6E 18 44  emotingVersion.D
000000a0  61 74 61 54 61 62 6C 65  2E 52 65 6D 6F 74 69 6E  ataTable.Remotin
000000b0  67 46 6F 72 6D 61 74 13  44 61 74 61 54 61 62 6C  gFormat.DataTabl
000000c0  65 2E 54 61 62 6C 65 4E  61 6D 65 13 44 61 74 61  e.TableName.Data
000000d0  54 61 62 6C 65 2E 4E 61  6D 65 73 70 61 63 65 10  Table.Namespace.
000000e0  44 61 74 61 54 61 62 6C  65 2E 50 72 65 66 69 78  DataTable.Prefix
000000f0  17 44 61 74 61 54 61 62  6C 65 2E 43 61 73 65 53  .DataTable.CaseS
00000100  65 6E 73 69 74 69 76 65  1E 44 61 74 61 54 61 62  ensitive.DataTab
00000110  6C 65 2E 63 61 73 65 53  65 6E 73 69 74 69 76 65  le.caseSensitive
00000120  41 6D 62 69 65 6E 74 14  44 61 74 61 54 61 62 6C  Ambient.DataTabl
00000130  65 2E 4C 6F 63 61 6C 65  4C 43 49 44 19 44 61 74  e.LocaleLCID.Dat
00000140  61 54 61 62 6C 65 2E 4D  69 6E 69 6D 75 6D 43 61  aTable.MinimumCa
00000150  70 61 63 69 74 79 19 44  61 74 61 54 61 62 6C 65  pacity.DataTable
00000160  2E 4E 65 73 74 65 64 49  6E 44 61 74 61 53 65 74  .NestedInDataSet
00000170  12 44 61 74 61 54 61 62  6C 65 2E 54 79 70 65 4E  .DataTable.TypeN
00000180  61 6D 65 1B 44 61 74 61  54 61 62 6C 65 2E 52 65  ame.DataTable.Re
00000190  70 65 61 74 61 62 6C 65  45 6C 65 6D 65 6E 74 1C  peatableElement.
000001a0  44 61 74 61 54 61 62 6C  65 2E 45 78 74 65 6E 64  DataTable.Extend
000001b0  65 64 50 72 6F 70 65 72  74 69 65 73 17 44 61 74  edProperties.Dat
000001c0  61 54 61 62 6C 65 2E 43  6F 6C 75 6D 6E 73 2E 43  aTable.Columns.C
000001d0  6F 75 6E 74 21 44 61 74  61 54 61 62 6C 65 2E 44  ount!DataTable.D
000001e0  61 74 61 43 6F 6C 75 6D  6E 5F 30 2E 43 6F 6C 75  ataColumn_0.Colu
000001f0  6D 6E 4E 61 6D 65 20 44  61 74 61 54 61 62 6C 65  mnName DataTable
```

图 7-6　序列化一个 DataSet 的结果

7.4.4　将 DataSet/DataTable 对象序列化为 JSON

下面介绍的 3 种方法可以将 DataTable 序列化为 JSON，第(1)、(2)种方法可以参考 5.5 节中的相关介绍。

1. 使用 Json. Net DLL(Newtonsoft)

使用 Json. NET 可以直接将表序列化为 JSON，下面是主要代码：

```
public string DataTableToJsonWithJsonNet(DataTable table)
{
  string JsonString = string.Empty;
  JsonString = JsonConvert.SerializeObject(table);
  return JsonString;
}
```

2. 使用 JavaScriptSerializer

JavaScriptSerializer 类是由异步通信层内部使用，用于序列化和反序列化在浏览器和 Web 服务器之间传递的数据。如果序列化成一个对象，就使用序列化方法；如果反序列化成 JSON 字符串，就使用 Deserialize 或 DeserializeObject 方法。下面的代码使用序列化方法得到 JSON 格式的数据：

```
public string DataTableToJsonWithJavaScriptSerializer(DataTable table)
{
```

```
        JavaScriptSerializer jsSerializer = new JavaScriptSerializer();
    List < Dictionary < string, object >> parentRow = new List < Dictionary < string, object >> ();
        Dictionary < string, object > childRow;
        foreach(DataRow row in table.Rows)
        {
            childRow = new Dictionary < string, object > ();
            foreach(DataColumn col in table.Columns)
            {
                childRow.Add(col.ColumnName, row[col]);
            }
            parentRow.Add(childRow);
        }
        return jsSerializer.Serialize(parentRow);
}
```

3. 使用 StringBuilder

输出 JSON，可满足指定格式的要求：

```
public string DataTableToJsonWithStringBuilder(DataTable dt)
{
    //由表转换为 json
    StringBuilder jsonBuilder = new StringBuilder();
    jsonBuilder.Append("{\"total\"" + ":" + dt.Rows.Count + ",\"rows\":[");
    for (int i = 0; i < dt.Rows.Count; i++)
    {
        jsonBuilder.Append("{");
        for (int j = 0; j < dt.Columns.Count; j++)
        {
            jsonBuilder.Append("\"");
            jsonBuilder.Append(dt.Columns[j].ColumnName);
            jsonBuilder.Append("\":\"");
            jsonBuilder.Append(dt.Rows[i][j].ToString());
            jsonBuilder.Append("\",");
        }
        if (dt.Columns.Count > 0)
        {
            jsonBuilder.Remove(jsonBuilder.Length - 1, 1);
        }
        jsonBuilder.Append("},");
    }
    if (dt.Rows.Count > 0)
    {
        jsonBuilder.Remove(jsonBuilder.Length - 1, 1);
    }
    jsonBuilder.Append("]}");
    return jsonBuilder.ToString();
}
```

使用 StringBuilder 编写输出 JSON 的格式与第(1)、(2)种方法稍有不同，它是 jQuery EasyUI 中 DataGrid 组件指定的 JSON 格式。

例 7-6　分别使用上述 3 种方法将 student 表序列化为 JSON 的格式输出。

❶ 引入以下命名空间：

```
using System.Data;
using System.Data.SqlClient;
using Newtonsoft.Json;
using System.Web.Configuration;
using System.Web.Script.Serialization;
using System.Text;
```

❷ 新建 Web 窗体 7_6.aspx，页面上添加 3 个命令按钮，即 Button1、Button2、Button3，设置 3 个按钮的标题。

❸ 将前面的 3 种方法加入到窗体中。

❹ 编写以下代码，返回 DataTable：

```
private DataTable GetDataTable()
{
  //从 Web.config 的 ConnectionStrings 节读取数据库连接字符串
  string cnnString = WebConfigurationManager.ConnectionStrings["SQLServerString"].ConnectionString;
  SqlConnection conn = new SqlConnection(cnnString);
  string strSQL = "select student_no as 学号, student_name as 姓名, gender as 性别 from student";
  SqlDataAdapter adapter = new SqlDataAdapter(strSQL, conn);
  //填充数据集 ds，将加入的表命名为 student
  DataSet ds = new DataSet();
  adapter.Fill(ds, "student");
  DataTable dt = ds.Tables[0];
  return dt;
}
```

❺ 在 Button1、Button2、Button3 的 Click 事件中，分别调用以上 3 种序列化 JSON 的方法如下：

```
protected void Button1_Click(object sender, EventArgs e)
{
  Response.Write(DataTableToJsonWithJsonNet(GetDataTable()));
}
protected void Button2_Click(object sender, EventArgs e)
{
  Response.Write(DataTableToJsonWithJavaScriptSerializer(GetDataTable()));
}
protected void Button3_Click(object sender, EventArgs e)
{
   Response.Write(DataTableToJsonWithStringBuilder(GetDataTable()));
}
```

❻ 运行 7_6.aspx 文件，单击 Button1 按钮和 Button2 按钮后的运行结果一样，如图 7-7 所示；单击 Button3 按钮后的运行结果如图 7-8 所示。

[{"学号":"30012035","姓名":"楚云飞","性别":"男"},{"学号":"40012030","姓名":"赵三","性别":"男"},{"学号":"41321059","姓名":"李孝诚","性别":"男"},{"学号":"41340136","姓名":"马小玉","性别":"女"},{"学号":"41355062","姓名":"王长林","性别":"男"},{"学号":"41361045","姓名":"李将寿","性别":"男"},{"学号":"41361258","姓名":"刘登山","性别":"男"},{"学号":"41361260","姓名":"张文丽","性别":"女"},{"学号":"41401007","姓名":"鲁宇星","性别":"男"},{"学号":"41405002","姓名":"王小月","性别":"女"},{"学号":"41405003","姓名":"张晨露","性别":"女"},{"学号":"41405005","姓名":"张伟达","性别":"男"},{"学号":"41405221","姓名":"张康林","性别":"女"},{"学号":"41405238","姓名":"张雨","性别":"女"},{"学号":"41405239","姓名":"段誉","性别":"男"}]

图 7-7　表序列化为 JSON 时使用方法 1 和方法 2 的输出结果

{"total":15,"rows":[{"学号":"30012035","姓名":"楚云飞","性别":"男"},{"学号":"40012030","姓名":"赵三","性别":"男"},{"学号":"41321059","姓名":"李孝诚","性别":"男"},{"学号":"41340136","姓名":"马小玉","性别":"女"},{"学号":"41355062","姓名":"王长林","性别":"男"},{"学号":"41361045","姓名":"李将寿","性别":"男"},{"学号":"41361258","姓名":"刘登山","性别":"男"},{"学号":"41361260","姓名":"张文丽","性别":"女"},{"学号":"41401007","姓名":"鲁宇星","性别":"男"},{"学号":"41405002","姓名":"王小月","性别":"女"},{"学号":"41405003","姓名":"张晨露","性别":"女"},{"学号":"41405005","姓名":"张伟达","性别":"男"},{"学号":"41405221","姓名":"张康林","性别":"女"},{"学号":"41405238","姓名":"张雨","性别":"女"},{"学号":"41405239","姓名":"段誉","性别":"男"}]}

图 7-8　表序列化为 JSON 时使用方法 3 的输出结果

7.4.5　DataSet 充填多个表和关系

例 7-7　student 表和 score 表是 1∶m 关系，编程将这两个表及其关系充填到 DataSet 中，程序运行后的效果如图 7-9 所示。选择 student 表中某行，可显示对应的 score 中的记录。

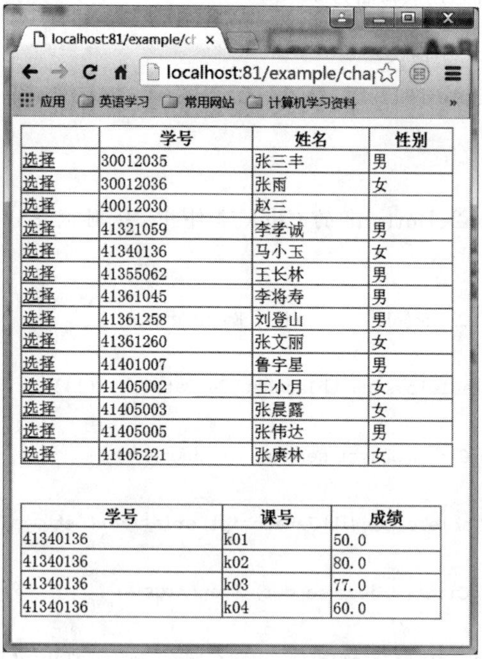

	学号	姓名	性别
选择	30012035	张三丰	男
选择	30012036	张雨	女
选择	40012030	赵三	
选择	41321059	李孝诚	男
选择	41340136	马小玉	女
选择	41355062	王长林	男
选择	41361045	李将寿	男
选择	41361258	刘登山	男
选择	41361260	张文丽	女
选择	41401007	鲁宇星	男
选择	41405002	王小月	女
选择	41405003	张晨露	女
选择	41405005	张伟达	男
选择	41405221	张康林	女

学号	课号	成绩
41340136	k01	50.0
41340136	k02	80.0
41340136	k03	77.0
41340136	k04	60.0

图 7-9　主表 student 与子表 score 的 1∶m 显示

❶ 新建 7_7.aspx 文件，页面中放置 GridView1 和 GridView2 两个控件。类中定义变量如下：

```
DataSet ds = new DataSet();
DataRelation relation;
```

❷ 在 Page_Load 中输入以下代码：

```
string cnnString = WebConfigurationManager.ConnectionStrings["SQLServerString"].ConnectionString;
SqlConnection conn = new SqlConnection(cnnString);
string strSQL = "select student_no as 学号,student_name as 姓名,gender as 性别 from student";
SqlDataAdapter adapter = new SqlDataAdapter(strSQL, conn);
//填充数据集 ds,将加入的表命名为 student
adapter.Fill(ds, "student");
//更改 DataAdapter 填充数据集的数据源
adapter.SelectCommand.CommandText = "select student_no as 学号,course_no as 课号,score as 成绩 from score";
//再次填充数据集 ds,将加入的表命名为 score
adapter.Fill(ds, "score");
//建立主表 student 和子表 score 的关系
relation = new DataRelation("student_score", ds.Tables["student"].Columns["学号"], ds.Tables["score"].Columns["学号"]);
ds.Relations.Add(relation);
GridView1.DataSource = ds.Tables["student"];
GridView1.DataBind();
```

说明：

① 每个 DataAdapter 对象都支持 4 个 Command 对象，分别是 InsertCommand、UpdateCommand、DeleteCommand 和 SelectCommand，但只有 SelectCommand 是填充 DataSet 所必需的。使用 DataAdapter 时可以先创建所需要的 Command 对象，然后将其赋值给 DataAdapter.SelectCommand 属性。在 DataAdapter 构造函数中，只要提供 Connection 对象和查询字符串就可以了。例如：

```
SqlDataAdapter adapter = new SqlDataAdapter(strSQL, conn);
```

② 关系通过定义 DataRelation 对象并将其加入到 DataRelations 集合来创建。创建 DataRelation 需要 3 个参数，即关系的名称、父表中作为主键的 DataColumn、子表中作为主键的 DataColumn。需要注意的是，如果填充数据源的 select 语句中使用了"select 字段名 as 列的别名……"，定义 DataRelation 时要使用这些别名，而不能使用列原来的名字。

❸ GridView1（显示主表 student）的 SelectedIndexChanged（表示选中的行发生改变后）事件中的代码如下：

```
//得到选中的数据行
DataRow studentRow = ds.Tables["student"].Rows[GridView1.SelectedIndex];
//得到与选中数据行相关联的子数据行
DataRow[] childRows = studentRow.GetChildRows(relation);
DataTable dt = new DataTable();
//建立 DataColumn 列,并设置名称为"学号"、类型为"字符串"
DataColumn studentColumn = new DataColumn("学号",typeof(string));
//将定义的列加到表中
dt.Columns.Add(studentColumn);
//定义"课号"列
```

```
studentColumn = new DataColumn("课号", typeof(string));
dt.Columns.Add(studentColumn);
//定义"成绩"列,数据类型为"小数"
studentColumn = new DataColumn("成绩", typeof(decimal));
dt.Columns.Add(studentColumn);
//循环子数据行
foreach (DataRow childRow in childRows)
{
    //为表 dt 建立新的数据行
    DataRow newRow = dt.NewRow();
    newRow["学号"] = childRow[0].ToString();
    newRow[1] = childRow[1];
    newRow[2] = childRow[2];
    dt.Rows.Add(newRow);
}
GridView2.DataSource = dt;
GridView2.DataBind();
```

说明:

① 程序中先得到主表 student 被选中的数据行 studentRow,然后通过 studentRow.GetChildRows(relation)获得与主表相关联子表的数据行,接着建立数据表 dt,将子表所有数据行的数据加到 dt,最后显示在 GridView2 中。

② 显示主表和子表的记录,也可通过带有 join 的 select 查询完成。什么时间使用 DataRow.GetChildRows()这种方式,取决于是否要更新获得的数据。如果要更新,使用独立的表和 DataRelation 对象通过灵活性会更大些;如果不需要更新,则两种方式都可以使用,但由于 join 查询只需要在网络上往返一次,所以查询效率会更高些。

7.4.6 利用 DataAdapter 更新数据库中的数据

使用 DataAdapter 更新数据库主要使用该类封闭的 InsertCommand、UpdateCommand、DeleteCommand 这 3 个命令实现。

例 7-8 使用 DataAdapter 对 student 表中记录更新、增加。

```
string cnnString = WebConfigurationManager.ConnectionStrings["SQLServerString"].ConnectionString;
SqlConnection conn = new SqlConnection(cnnString);
string strSQL = "select student_no,student_name from student";
SqlDataAdapter adapter = new SqlDataAdapter(strSQL, conn);
DataSet ds = new DataSet();
adapter.Fill(ds,"student");
DataTable dt = ds.Tables["student"];
GridView1.DataSource = dt;
GridView1.DataBind();
GridView1.Caption = "original data";
//在内存中对 DataTable 进行操作
//增加一条记录
DataRow newRow = dt.NewRow();
newRow["student_no"] = "A9401843";
newRow["student_name"] = "赵飞强";
```

```
dt.Rows.Add(newRow);
//更新第 1 条记录的姓名
newRow = dt.Rows[0];
newRow["student_name"] = "楚云飞";
//将内存中的表数据变化更新到数据库中
SqlCommandBuilder cmdBuilder = new SqlCommandBuilder(adapter);
adapter.Update(ds, "student");
GridView2.DataSource = dt;
GridView2.DataBind();
GridView2.Caption = "updated data";
```

说明：

① 更新数据库中数据时，使用了 SqlCommandBuilder 类，该类的对象可以让系统自动生成对应的 Insert、Update 和 Delete 操作的命令。

② 调用 DataAdapter 的 Update 方法，以执行自动生成的 3 个命令对象，从而更新数据库。

7.5　DataView 类

DataView 就是用来表示排序、筛选、搜索、编辑和导航的 DataTable 的可绑定数据的自定义视图。利用 DataAdapter 将服务器上数据填充到 DataSet 后，与服务器断开连接。DataView 提供了基于 DataTable 数据的动态视图。

例 7-9　利用 DataView 根据性别显示 student 中数据，程序运行结果如图 7-10 所示。单击“男”按钮显示男的记录，单击“女”按钮显示女的记录，单击“全部”按钮显示全部记录。

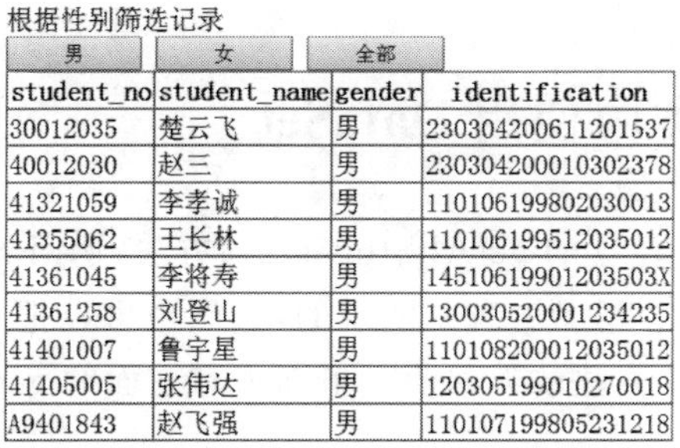

根据性别筛选记录

男　女　全部

student_no	student_name	gender	identification
30012035	楚云飞	男	230304200611201537
40012030	赵三	男	230304200010302378
41321059	李孝诚	男	110106199802030013
41355062	王长林	男	110106199512035012
41361045	李将寿	男	14510619901203503X
41361258	刘登山	男	130030520001234235
41401007	鲁宇星	男	110108200012035012
41405005	张伟达	男	120305199010270018
A9401843	赵飞强	男	110107199805231218

图 7-10　DataView 筛选记录

❶ 定义 Web 页面的变量 DataView dv;，页面上放置 GridView1 和 3 个命令按钮，设置 3 个按钮的 Text 属性。

❷ 在 Page_Load 中输入以下代码：

```
string cnnString = WebConfigurationManager.ConnectionStrings["SQLServerString"].ConnectionString;
SqlConnection conn = new SqlConnection(cnnString);
string strSQL = "select student_no,student_name,gender,identification from student";
SqlDataAdapter adapter = new SqlDataAdapter(strSQL, conn);
DataSet ds = new DataSet();
adapter.Fill(ds, "student");
DataTable dt = ds.Tables["student"];
dv = new DataView(dt);
```

❸ 在表示"男"的命令按钮的 Click 事件中输入以下代码：

```
dv.RowFilter = "gender = '男'";
GridView1.DataSource = dv;
GridView1.DataBind();
```

❹ 在表示"女"的命令按钮的 Click 事件中输入以下代码：

```
dv.RowFilter = "gender = '女'";
GridView1.DataSource = dv;
GridView1.DataBind();
```

❺ 在表示"全部"的命令按钮的 Click 事件中输入以下代码：

```
dv.RowFilter = "";
GridView1.DataSource = dv;
GridView1.DataBind();
```

说明：

① DataView 过滤使用 RowFilter 属性设置过滤条件。

② 如果要使用 DataView 排序，将 Sort 属性设置为有效的排序表达式就可以了，如例中要使用 student_no 排序，语句为

```
dv.Sort = "student_no";
```

如果要使用多个字段排序，字段间用逗号分隔，如：

```
dv.Sort = "student_no,student_name";
```

7.6 在数据访问类中使用 DataSet

第 5 章在 DBbase 类中建立的 getAllStudents() 方法，返回 student 表中的记录，是通过使用 DataReader 和 List<>泛型的方式实现的，当然也可以把 DataSet 或者 DataTable 作为自定义数据访问类中某个方法的返回值。下面重写前面的 getAllStudents()：

```
public static DataTable getAllStudents()
{
    string sql = "select * from student";
    SqlDataAdapter adapter = new SqlDataAdapter(sql, conn);
    DataSet ds = new DataSet();
    adapter.Fill(ds);
    return ds.Tables[0];
}
```

7.7 编写与提供程序无关的程序代码

在有些情况下，程序开发完成后将来可能要移植到不同的数据库，或者无法确定应用程序的最终版本使用哪种数据库时，使用与提供程序无关的程序代码就非常有必要了；否则

需要不断地修改 ADO.NET 各对象的名称,非常麻烦,难以构建 ADO.NET 的工具或者插件。从.NET 2.0 开始引进了新的工厂模型,增强了对提供程序无关代码的支持。

7.7.1 创建工厂

工厂模型的基本思想是借助单一的工厂模型创建出所有需要使用的提供程序相关的对象。

.NET 提供了一个工厂的标准类 System.Data.Common.DbProviderFactories,该类有一个静态的 GetFactory()方法,这个方法根据不同的提供程序的名字,能够返回相应的工厂。

下面是使用 DbProviderFactories 建立的 SqlClientFactory 工厂(访问 SQL Server):

```
String factory = "System.Data.SqlClient";
DbProviderFactory provider = DbProviderFactories.GetFactory(factory);
```

使用 DbProviderFactories 建立的 OleDbFactory 工厂(可访问 Access 数据库):

```
String factory = "System.Data.OleDb";
DbProviderFactory provider = DbProviderFactories.GetFactory(factory);
```

使用 DbProviderFactories 建立的 OracleClientFactory 工厂(可访问 Oracle 数据库):

```
String factory = "System.Data.OracleClient";
DbProviderFactory provider = DbProviderFactories.GetFactory(factory);
```

从前面的代码可以看出,只要给 DbProviderFactories.GetFactory()方法传送标识提供程序名称的字符串,就能得到访问不同数据库的工厂,为此可以将不同提供程序名称的字符串都写到 Web.config 中。下面针对 SQL Server 和 Access 在 Web.config 中注册工厂:

```
<connectionStrings>
<add name = "AccessString" connectionString = "Provider = Microsoft.ACE.OLEDB.12.0;data
source = |DataDirectory|students.accdb"/>
<add name = "SQLServerString" connectionString = "server = .;database = students;uid = sa;pwd =
3500"/>
</connectionStrings>
<appSettings>
    <add key = "SQLServer" value = "System.Data.SqlClient"/>
    <add key = "access" value = "System.Data.OleDb"/>
</appSettings>
</configuration>
```

7.7.2 使用工厂建立对象

在拥有一个工厂后,就可以使用 DbProviderFactory.Create××××()方法创建其他对象了,如建立 Connection 对象:

```
DbConnection connection = provider.CreateConnection();
```

建立 Command 对象:

```
DbCommand cmd = provider.CreateCommand();
```

表 7-4 列出了使用何种方法来创建数据库访问的对象。

表 7-4 标准 ADO.NET 对象接口

对象类型	基 类	示 例	DbProviderFactory 方法
Connection	DbConnection	SqlConnection	CreateConnection()
Command	DbCommand	SqlCommand	CreateCommand()
Parameter	DbParameter	SqlParameter	CreateParameter()
DataReader	DbDataReader	SqlDataReader	None（使用 IdbCommand.ExecuteReader()）
DataAdapter	DbDataAdapter	SqlDataAdapter	CreateDataAdapter()

7.7.3 使用与程序无关的代码查询示例

❶ 业务逻辑。建立一个 factroy.cs 类，代码如下：

```
1  using System;
2  using System.Collections.Generic;
3  using System.Linq;
4  using System.Web;
5  using System.Data.Common;
6  using System.Data;
7  using System.Configuration;
8  namespace aspExample.chapter5
9  {
10    public class factory
11    {
12      static string connectionString = ConfigurationManager.ConnectionStrings["AccessString"].ConnectionString;
13      static DbProviderFactory provider = DbProviderFactories.GetFactory(ConfigurationManager.AppSettings["access"]);
       #region 传入 SQL，返回 DataSet
14      public static DataSet GetDataSet(string SQLString)
15      {
16        using (DbConnection connection = provider.CreateConnection())
17        {
18          connection.ConnectionString = connectionString;
19          using (DbCommand cmd = provider.CreateCommand())
20          {
21            cmd.Connection = connection;
22            cmd.CommandText = SQLString;
23            try
24            {
25              DataSet ds = new DataSet();
26              DbDataAdapter adapter = provider.CreateDataAdapter();
27              adapter.SelectCommand = cmd;
28              adapter.Fill(ds, "ds");
29              return ds;
30            }
```

```csharp
31              catch (DbException ex)
32              {
33                  connection.Close();
34                  connection.Dispose();
35                  throw new Exception(ex.Message);
36              }
37          }
38      }
39  }
    #endregion
    #region 列出全部学生,返回泛型集合
40  public static List<studentData> GetStudents()
41  {
42      using (DbConnection connection = provider.CreateConnection())
43      {
44          connection.ConnectionString = connectionString;
45          connection.Open();
46          List<studentData> listStudents = new List<studentData>();
47          using (DbCommand cmd = provider.CreateCommand())
48          {
49              cmd.CommandText = "select * from student";
50              cmd.Connection = connection;
51              DbDataReader reader = cmd.ExecuteReader();
52              while (reader.Read())
53              {
54                  studentData students = new studentData();
55                  students.student_no = (string)reader["student_no"];
56                  students.student_name = (string)reader["student_name"];
57                  students.identification = (string)reader["identification"];
58                  students.id = (int)reader["id"];
59                  listStudents.Add(students);
60              }
61              reader.Close();
62          }
63          return listStudents;
64      }
65  }
    #endregion
66  }
67  }
68  }
```

说明:

① 第 12 行得到 Web.config 的 <connectionStrings> 节 name = "AccessString" 的 connectionString 的值,connectionString 的值是" Provider = Microsoft.ACE.OLEDB.12.0; data source = |DataDirectory|students.accdb",是 Access 数据库提供程序的名称,|DataDirectory|是 App_Data 文件夹,students.accdb 要放在该文件夹下。

② 第 13 行定义工厂 provider,为 GetFactoory()方法传送的字符串是从 Web.config 的 <appSettings> 节中读取的"System.Data.OleDb"。

如果要访问的是 SQL Server 数据库,需要将第 12 行的["AccessString"]改为

["SQLServerString"],第 13 行的["access"]改为["SQLServer"]即可。

③ 第 14～39 行定义一个方法 GetDataSet(),传入 select 语句,返回 DataSet。

④ 第 19 行用工厂建立 Command 对象 cmd,而 26 行用工厂建立 DataAdapter 对象 adapter。

⑤ 第 40～65 行定义方法 GetStudents(),以泛型集合的方式返回全部学生的记录。

❷ 用户界面。建立 Web 窗体 7_11.aspx,在窗体中增加 GridView1 控件,引入命名空间 using System.Data;后,在 Page_Load 事件中写入如下代码:

```
DataSet ds = new DataSet();
ds = factory.GetDataSet("select * from student");
GridView1.DataSource = ds;
GridView1.DataBind();
```

或者

```
List < studentData > students = new List < studentData >();
students = factory.GetStudents();
GridView1.DataSource = students;
GridView1.DataBind();
```

运行 7_11.aspx,前后两段代码的效果完全相同,只是后面使用了数据工厂的编程方式。

习题与思考

(1) 将 score 表二进制序列化,结果保存为 score.bin。

(2) 有一个名为 finishers.xml 的文件,记录运动员到达终点的时间,内容如下,用 DataSet 读取该 XML 文件,并显示在 GridView 中。

```
<?xml version = "1.0" encoding = "utf - 8"?>
<finishers>
    <runner>
        <fname>John</fname>
        <lname>Smith</lname>
        <gender>m</gender>
        <time>25:31</time>
    </runner>
    <runner>
        <fname>Mary</fname>
        <lname>Brown</lname>
        <gender>f</gender>
        <time>26:01</time>
    </runner>
    <runner>
        <fname>Frank</fname>
        <lname>Jones</lname>
        <gender>m</gender>
        <time>26:08</time>
```

```
        </runner>
</finishers>
```

(3) 通过工厂建立命令对象，从 App_Data 文件夹下的 students.accdb 库中删除 student 表中 student_no="50123456" 的记录，请填空。

Web.config 中的内容如下：

```
<connectionStrings>
  <add name = "AccessString" connectionString = _____ />
</connectionStrings>
<appSettings>
  <add key = "access" value = _____ />
</appSettings>
```

程序代码放在某个 aspx 的 Page_Load() 事件中：

```
protected void Page_Load(object sender, EventArgs e)
{
    string connectionString = ConfigurationManager.ConnectionStrings["AccessString"].ConnectionString;
    DbProviderFactory provider = DbProviderFactories.GetFactory(ConfigurationManager.AppSettings["access"]);
    using (DbConnection connection = _____)
    {
        connection.ConnectionString = connectionString;
        connection.Open();
        using (DbCommand cmd = _____)
        {
            cmd.CommandText = "_____";
            cmd.Connection = connection;
            cmd._____;
        }
    }
}
```

(4) 通过定义一个 Table 将表 7-5 中的记录增加到该 Table 中。

表 7-5　习题与思考题(4)表

字段名	studentID	courseID	score
值	20102	K1	90

根据表 7-5 的内容阅读代码并填空：

```
DataTable dt = new DataTable("score");
DataColumn column = _____("courseID", _____ );
_____;
column = _____ ("studentID", typeof(Int16));
_____;
column = _____ ("score", typeof(decimal));
_____;
```

```
                    _____;
dr["studentID"] = 20102;
dr["courseID"] = "k1";
dr["score"] = 90;
                    _____;
```

(5) 将 student 表中的学生姓名用 RadioButtonList 控制显示出来，每行列出 5 个学生的姓名，运行后的效果如图 7-11 所示。

提示：RadioButtonList 绑定数据时，除了要设置 DataSource 属性和 DataBind()，还需要设置 DataTextField 和 DataValueField 属性。

图 7-11 用 RadioButtonList 控件列出表中的姓名

(6) 读取 students 库中相关的表，查询 student_no、student_name、course_name、score.score 4 列，并显示在 GridView1 中。运行结果见图 7-12。

学号	姓名	课程名	成绩
41321059	李孝诚	ASP网络数据库	80.0
41321059	李孝诚	高等数学	90.0
41321059	李孝诚	英语	67.0
41340136	马小玉	ASP网络数据库	50.0
41340136	马小玉	高等数学	55.0
41340136	马小玉	英语	77.0
41340136	马小玉	人工智能	84.0
41355062	王长林	ASP网络数据库	65.0
41355062	王长林	高等数学	83.0
41355062	王长林	英语	79.0

图 7-12 用 DataGrid1 显示学生成绩

(7) 编写程序，将 teacher 表序列化为 XML。

第 8 章

ASP.NET内部控件

(1) GridView 控件。

(2) UpLoad 控件。

(3) Chart 控件。

(4) TreeView 控件。

ASP.NET 服务器控件是 ASP.NET 架构的基础组成部分,本质上服务器控件就是.NET Framework 中用来表示 Web 窗体中可见元素的类,一些类相对简单,直接与特定的 HTML 标记对应,而另一些类则是对来自多个 HTML 标记复杂表现形式更深的抽象。

服务器控件的类型包括有以下几种。

(1) HTML 服务器控件。这类控件封装了标准的 HTML 标记。在某个普通的 HTML 标记上添加 runat="server",就可以将该 HTML 标记转换成服务器控件。

(2) Web 控件。这些类复制了 HTML 标记的功能,有一组一致且富有含义的属性和方法,如 HyperLink、ListBox 和 Button 等控件。

(3) 富控件。这些控件能够生成大量的 HTML 标记,甚至可以生成客户端的 JavaScript 来创建用户界面,如 TreeView、Calendar 等控件。

(4) 验证控件。使用这些控件可以方便地按照几个标准或用户定义的规则去验证关联的输入控件。

(5) 数据控件。用来显示大量数据的复杂网格和列表,支持如模板、编辑、排序和分页等高级特性,这些控件也包括数据源控件。

(6) 导航控件。用于显示站点地图,允许用户由一个页面导航到另一个页面。

由于篇幅所限,本章只介绍 GridView、UpLoad、Chart、TreeView 这 4 个控件。

8.1 GridView 控件

ASP.NET 提供了许多控件来显示表格的数据,其中 GridView 是应用最多、功能最强的一个,通过该控件可以显示、编辑、删除多种不同的数据源(如数据库、XML、文件和公开数据的业务对象)中的数据,可以通过使用数据源控件自动绑定和显示数据,并对数据进行选择、排序、分页、编辑和删除。用户可以自定义 GridView 控件的外观,通过处理 GridView 的事件自定义该控件的行为。

例 8-1 使用 GridView 控件的自定义行为,显示、编辑、删除 student 表中的数据,运行效果如图 8-1 所示。页面支持分页,每页 5 条记录;单击"编辑"链接,显示如图 8-2 所示,数据修改后,单击"更新"链接完成数据的修改;单击"选择"链接,将选中的"姓名"放在文本框 txtName 中;单击"删除"链接,弹出"是否删除"对话框,确认删除后,删除指定的记录。

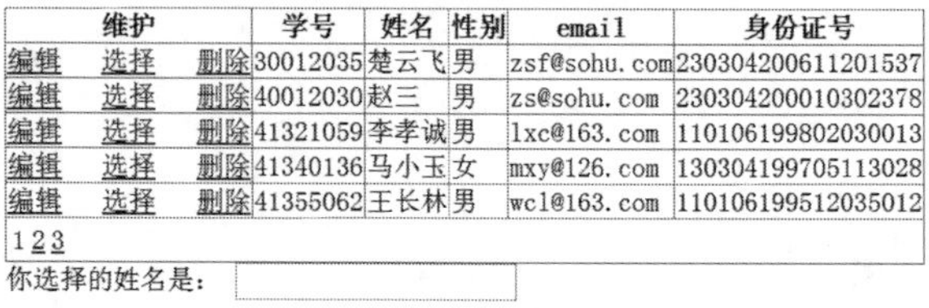

图 8-1 使用 GridView 控件显示数据

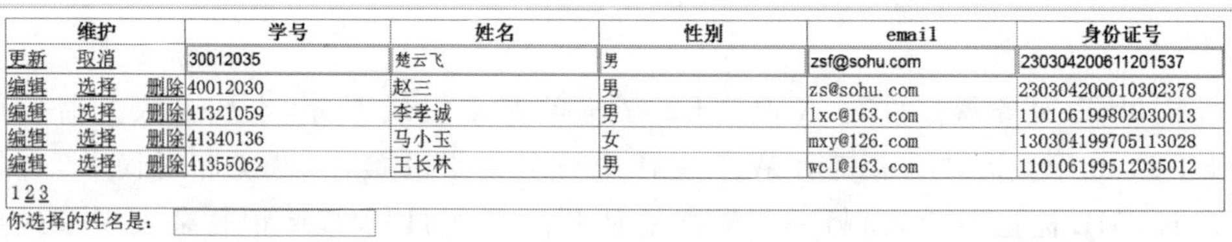

图 8-2 使用 GridView 控件编辑数据

❶ 新建 Web 窗体 8_1.aspx,在页面上放置一个 GridView 控件 GridView1、文本框控件 txtName 和一个标签控件,设置标签控件的 Text 属性值为"你选择的姓名是:"。

❷ 在页面的设计视图中选中 GridView1,右击控件右上角的小按钮,选择快捷菜单中的"编辑列"命令,如图 8-3 所示。

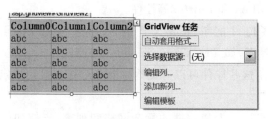

图 8-3 GridView 任务对话框

❸ 在弹出的"字段"对话框中,从"可用字段"中选中 BoundField 选项,单击"添加"按钮,设置新增加字段的 DataField 属性为 student_no,HeaderText 属性为"学号"(图 8-4);再增加 5 个新的字段,设置其 DataField 属性为 student_name、gender、email、identification、id,这些属性值要与 student 表中的字段保持一致,设置 HeaderText 属性为"姓名""性别"

"email""身份证号""id"。将 id 列的 Visible 设置为 false。

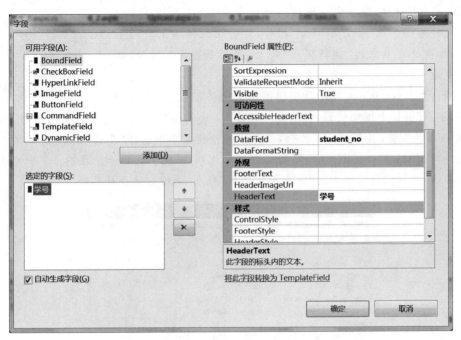

图 8-4　设置"字段"对话框 1

❹ 在"字段"对话框中，从"可用字段"中选中 CommandField 选项，单击"添加"按钮，设置该字段的 ShowDeleteButton、ShowEditButton、ShowSelectButton、ShowHeader 为 true，设置 HeaderText 为"维护"；通过单击"选定的字段"右侧的箭头按钮，将 CommandField 字段移动到最上面，如图 8-5 所示。

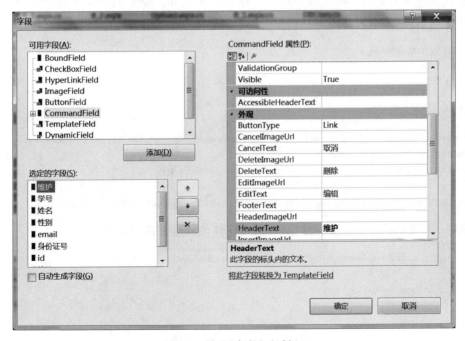

图 8-5　设置"字段"对话框 2

❺ 图 8-5 中选中"选定的字段"中的"维护"字段,然后单击右下方的"将此字段转换为 TemplateField"链接,即可完成模板的转换。单击"确定"按钮,回到窗体设计页面。选中 GridView1 控件,右击该控件右上角的小按钮,选择快捷菜单中的"编辑模板"命令,显示如图 8-6 所示,从图中可以看出,维护列是作为第 1 列出现的。

图 8-6 编辑模板列窗口

说明:由于 GridView 是微软提供的集成服务器控件,其不能直接在"删除"链接上增加代码,需要使用 GridView 的模板编辑功能来解决此问题。

❻ 在图 8-6 中,右击"删除"按钮,在弹出的快捷菜单中选择"属性"命令,进入右侧的控件属性编辑窗口,在 OnClientClick 属性中增加代码,如图 8-7 所示。在图 8-7 中右击右上角的小按钮,选择快捷菜单中的"结束模板编辑"命令,返回窗体设计界面。

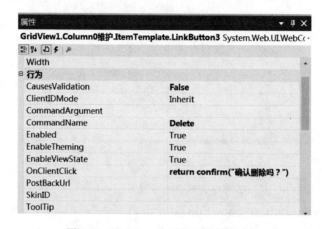

图 8-7 GridView 中删除按钮属性窗口

❼ 设置分页的属性:设置 GridView1 的 AllowPaging 属性为 True,PageSize 为 5。

❽ 双击页面,进入到代码窗口,引入命名空间:

```
using System.Data;
using System.Data.SqlClient;
using System.Web.Configuration;
```

❾ 编写以下代码:

```
static string cnnString = WebConfigurationManager.ConnectionStrings["SQLServerString"].ConnectionString;
static SqlConnection conn = new SqlConnection(cnnString);
static DataSet ds = new DataSet();
SqlDataAdapter adapter = new SqlDataAdapter("select * from student", conn);
```

说明:上述代码需要放在窗体中,以便在窗体的其他事件中使用。

```
protected void Page_Load(object sender, EventArgs e)
{
    if (!Page.IsPostBack)
    {
        bind();;
    }
}
```

```csharp
}
//绑定数据
public void bind()
{
    ds.Clear();
    adapter.Fill(ds, "student");
    GridView1.DataSource = ds;                          //指定数据源
    GridView1.DataBind();                               //绑定数据
    GridView1.DataKeyNames = new string[] { "id" };     //设置主键
}
```

说明：GridView1.DataKeyNames = new string[] {"id"};设置访问 GridView1 每行的主键是 id。主键对应数据表中的一个或者多个字段。当主键是多个字段时，用逗号分隔字段名列表。GridView1 中 id 列是隐藏列，设置其主键后，在执行更新、删除等时，可取出此键，如本示例中更新记录时，取出此键值的语句：

```csharp
int id = Convert.ToInt16(GridView1.DataKeys[e.RowIndex].Values["id"]);
//编辑记录
protected void GridView1_RowEditing(object sender, GridViewEditEventArgs e)
{
    GridView1.EditIndex = e.NewEditIndex;
    bind();
}
//取消编辑
protected void GridView1_RowCancelingEdit(object sender, GridViewCancelEditEventArgs e)
{
    //GridView 都是从 0 行开始的，-1 表示取消编辑状态
    GridView1.EditIndex = -1;
    bind();
}
//更新记录
protected void GridView1_RowUpdating(object sender, GridViewUpdateEventArgs e)
{
    string student_no = ((TextBox)(GridView1.Rows[e.RowIndex].Cells[1].Controls[0])).Text.ToString().Trim();
    string student_name = ((TextBox)(GridView1.Rows[e.RowIndex].Cells[2].Controls[0])).Text.ToString().Trim();
    string gender = ((TextBox)(GridView1.Rows[e.RowIndex].Cells[3].Controls[0])).Text.ToString().Trim();
    string email = ((TextBox)(GridView1.Rows[e.RowIndex].Cells[4].Controls[0])).Text.ToString().Trim();
    string identification = ((TextBox)(GridView1.Rows[e.RowIndex].Cells[5].Controls[0])).Text.ToString().Trim();
    int id = Convert.ToInt16( GridView1.DataKeys[e.RowIndex].Values["id"]);
    string sqlstr = string.Format("update student set student_no = '{0}',student_name = '{1}',gender = '{2}',email = '{3}',identification = '{4}' where id = {5}", student_no, student_name, gender, email, identification, id);
    SqlCommand sqlcom = new SqlCommand(sqlstr, conn);
    conn.Open();
    sqlcom.ExecuteNonQuery();
```

```
        conn.Close();
        GridView1.EditIndex = -1;
        bind();
    }
    //单击"选择"链接时,在 txtName 文本框中显示选中行中的"姓名"
    protected void GridView1_SelectedIndexChanging(object sender, GridViewSelectEventArgs e)
    {
        txtName.Text = GridView1.Rows[e.NewSelectedIndex].Cells[2].Text;
    }
    //删除记录
    protected void GridView1_RowDeleting(object sender, GridViewDeleteEventArgs e)
    {
        string student_no = GridView1.Rows[e.RowIndex].Cells[1].Text;
        string sqlstr = string.Format("delete from student where student_no = '{0}'", student_no);
        SqlCommand sqlcom = new SqlCommand(sqlstr, conn);
        conn.Open();
        sqlcom.ExecuteNonQuery();
        conn.Close();
        GridView1.EditIndex = -1;
        bind();
    }
    //换页
    protected void GridView1_PageIndexChanging(object sender, GridViewPageEventArgs e)
    {
        GridView1.PageIndex = e.NewPageIndex;
        bind();
    }
```

8.2 Upload 控件

ASP.NET 中通常使用 FileUpload Web 服务器控件,将客户端文件上传至服务器。FileUpload 控件可以上传图片、文本文件或者其他文件。

例 8-2 上传一个图片,并显示在 Image 中。

❶ 新建 Web 窗体 8_2.aspx,在页面上拖放 FileUpload 控件、命令按钮和 Image 控件,页面布局如图 8-8 所示。在 8_2.aspx 文件所在的文件夹下建立一个文件夹 uploadFiles。

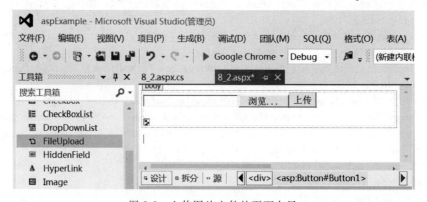

图 8-8 上传图片文件的页面布局

❷ 单击"上传"命令按钮,写入以下代码:

```
1  protected void Button1_Click(object sender, EventArgs e)
2  {
3      string fileName = FileUpload1.PostedFile.FileName;
4      if (fileName == "") return;
5      if (fileName.Substring(fileName.Length - 3, 3).ToLower() == "jpg")
6      {
7          fileName = Server.MapPath("uploadFiles/" + fileName);
8          FileUpload1.PostedFile.SaveAs(filename);
9          Image1.ImageUrl = "uploadFiles/" + fileName;
10     }
11 }
```

说明:
① 第 3 行获得要上传的文件名。
② 第 5 行判断要上传的文件名的扩展名是否为 jpg,本示例只允许上传 jpg 的图像文件。
③ 第 7 行将上传的文件转变为存放在服务器上的绝对地址;服务器上当前网页所在文件夹下要先建立一个名为 uploadFiles 的文件夹,上传后的文件存放在该文件夹下。
④ 第 8 行将要上传的文件保存到服务器。
上传文件时,如果 FileUpload 控件要限制上传文件的大小、类型等,需要使用以下属性。
① 获取上传文件的大小(单位 byte):

int fileSize = FileUpload1.PostedFile.ContentLength;

② 获取上传文件的扩展名:

string fileName = FileUpload1.PostedFile.FileName;
string exeName = fileName.Substring(filename.LastIndexOf(".")).ToLower();

8.3 Chart 控件

Chart 控件是 VS2012 中新增加的一个图表控件,使用该控件可以实现直方图、曲线走势图、饼状比例图等,甚至可以是混合图表,可以是二维或三维图表,可以带或不带坐标系,自由配置各条目的颜色、字体等。Chart 控件的结构如图 8-9 所示。

从图 8-9 中可以看出,Chart 可以绘制多个 ChartArea,每个 ChartArea 都可以绘制多条 Series。ChartArea 就是绘图区域,可以有多个 ChartArea 叠加在一起,Series 是画在 ChartArea 上的,Series 英文意思是"序列、连续",其实就是数据线,它可以是曲线、点、柱形、条形、饼图等图形。

为 Chart 控件添加数据,可以采用以下方式。
(1) 在 Chart 的设计界面中,在属性窗口的 Series 属性下的 Points 中添加需要的数据。
(2) 在 Chart 的设计界面中,在属性中绑定一个数据源。
(3) 通过程序代码,动态添加数据到 Chart 中。

(4)设置一个或多个数据源,直接绑定到 Chart 中。

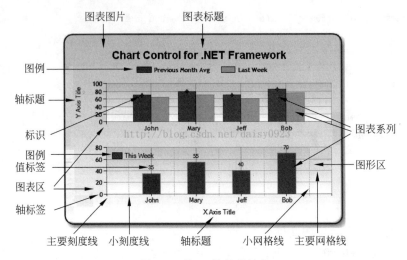

图 8-9　Chart 控件的结构

本章主要讲述第(3)、(4)种方法,通过程序代码,添加数据或者绑定数据到 Chart 控件。示例中需要引入的命名空间(字体加黑的是在 VS2012 中需要手工引入的命名空间)如下:

```
using System;
using System.Collections.Generic;
using System.Linq;
using System.Web;
using System.Web.UI;
using System.Web.UI.WebControls;
using System.Web.UI.DataVisualization.Charting;
using System.Data;
using System.Data.SqlClient;
using System.Drawing;
```

需要注意的是,Chart 控件以 IIS 方式(不是直接在 VS 中)运行,在浏览器中要正确地运行,需要在 Web.config 中增加以下内容:

```
<appSettings>
  <add key="ChartImageHandler" value="storage=file;timeout=20;Url=~/TempImageFiles/;" />
</appSettings>
```

需要在 Web.config 所在文件夹下建立一个与上面 URL 指定的同名的文件夹,如 TempImageFiles,用于处理 Chart 控件生成的图形。

如果上传到云上,则需要将上面的内容调整为:

```
<appSettings>
  <add key="ChartImageHandler" value="storage=file;timeout=20" />
</appSettings>
```

需要设置程序所在的文件夹具有读写权限,程序运行后会在 Web.config 所在文件夹下创建一个 charts_0 文件夹,存储 Chart 控件生成的临时文件。

8.3.1 Chart 控件添加数据

添加数据的方式相对较简单,为 Chart.Series["序列名"或者编号].Points 使用 Add、AddXY、AddY 等方法即可。

例 8-3 绘制一条正弦曲线和余弦曲线,效果如图 8-10 所示。

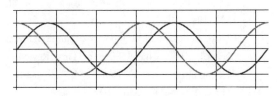

图 8-10 正弦曲线和余弦曲线

```
Chart1.Series.Clear();
Chart1.Series.Add("serieA");
Chart1.Series.Add("serieB");
double t;
for (t = (-2 * Math.PI); t <= (2 * Math.PI); t += Math.PI / 100)
{
    double ch1 = Math.Sin(t);
    double ch2 = Math.Cos(t);
    Chart1.Series["serieA"].ChartType = SeriesChartType.Spline;
    Chart1.Series["serieB"].ChartType = SeriesChartType.Spline;
    Chart1.Series["serieA"].Points.AddXY(t,ch1);
    Chart1.Series["serieB"].Points.AddXY(t,ch2);
}
```

说明:本程序为 Chart 的两个数据序列 serieA 和 serieB,使用 AddXY 的方法增加数据。Series 的 ChartType 属性指定绘图的类型。

8.3.2 Chart 控件数据绑定

能够为 Chart 提供数据源的类型有 SqlCommand、OleDbCommand、SqlDataAdapter、OleDbDataAdapter、DataView、DataSet、DataRow、DataReader、List、Array、IList、IListSource、IEnumerable,程序开发中最常用的是 DataView、DataReader、DataSet、DataRow、Array、List 这几种类型。图表控件的绑定方式有两种,即常规绑定和交叉表绑定。图表控件的 Y 轴数据,有的方法支持一次绑定多个值,以绘制时间、区域、使用量等之类的图形。Chart 控件绑定数据的主要方法见表 8-1。

表 8-1 Chart 控件绑定数据的主要方法

绑定的方法	功 能 说 明
Chart.DataBindTable	(1) 简单地绑定 X、Y 的值,根据数据源中列的数量,自动创建数据序列 (2) 每个数据序列不支持多个 Y 值,所有数据序列有相同的 X 值,或者不需要设置 X (3) 不支持绑定如 ToolTips 等图表扩展的属性

续表

绑定的方法	功 能 说 明
Chart. DataSource 和 Chart. DataBind	(1) 在设计视图中可以使用这两个方法 (2) 可以绑定多个 Y 值 (3) 不支持绑定如 ToolTips 等图表扩展的属性
Chart. Points. DataBind(X)Y	(1) 除了具有上述三者的功能外,支持绑定如 ToolTips 等图表扩展的属性,如: Chart1. Series["Series1"]. Points. DataBind(myDs. Tables[0]. DefaultView, "Name", "ID", "Label=ID,ToolTip=RegionID"); 其中 myDs 是 DataSet,"Name"、"ID"、RegionID 是 DataSet 中数据表中的字段名,"Name"是 X 轴,"ID"是 Y 轴 (2) 同一个数据序列,X、Y 的值必须是相同的数据源,如: Chart1. Series["Series1"]. Points. DataBindXY(myReader, "Name", myReader, "Sales"); 其中,"Name"为 X 轴,"Sales"为 Y 轴,myReader 是 DataReader
Chart. DataBindCrossTab	交叉数据的绑定,它可以根据数据动态地生成数据序列

如果要对数据序列 Series1 的 X 轴、Y 轴绑定某个字段,也可以采用以下代码:

```
//设置数据源,myDv 是一个取出数据集的 DataView
Chart1.DataSource = myDv;
//分别设置图表的 X 值和 Y 值
Chart1.Series["Series1"].XValueMember = "Name";
Chart1.Series["Series1"].YValueMembers = "Sales";
//绑定设置的数据
Chart1.DataBind();
```

例 8-4 统计 student 表中的男、女人数,用直方图显示,效果如图 8-11 所示。

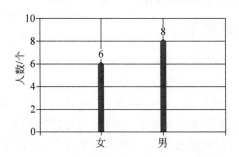

图 8-11 使用 DataBindTable 方法显示"性别"分布

引入相关的命名空间后,实现的代码如下:

```
string cnnString = WebConfigurationManager.ConnectionStrings["SQLServerString"].ConnectionString;
SqlConnection conn = new SqlConnection(cnnString);
string strSQL = "select gender as 性别,count(*) as 人数 from student group by gender order by 性别 desc";
SqlCommand cmd = new SqlCommand(strSQL, conn);
conn.Open();
SqlDataReader dr = cmd.ExecuteReader();
```

```
Chart1.ChartAreas[0].AxisX.LineColor = Color.Blue;        //X 轴颜色
Chart1.ChartAreas[0].AxisY.LineColor = Color.Blue;        //Y 轴颜色
Chart1.ChartAreas[0].AxisX.LineWidth = 2;                 //X 轴宽度
Chart1.ChartAreas[0].AxisY.LineWidth = 2;                 //Y 轴宽度
Chart1.ChartAreas[0].AxisY.Title = "人数";                //Y 轴标题
Chart1.Series.Clear();                                    //清除所有 Series
//DataBindTable 方法自动为 Chart 增加一个 Series,Series 的编号是 0
Chart1.DataBindTable(dr, "性别");                         //dr 必须处于打开状态
//Chart1.Series.Add("Series1");                           //如果增加此语句,Series1 的 Series 编号是 1
//下面注释的语句用于指定 Series1 的绘图区域,Chart 默认有一个 ChartArea
//Chart1.Series["Series1"].ChartArea = Chart1.ChartAreas[0].Name;
Chart1.Width = 350;
Chart1.Series[0]["PointWidth"] = "0.1";                   //设置柱的宽度
Chart1.Series[0].IsValueShownAsLabel = true;              //是否显示数据
Chart1.Series[0].IsVisibleInLegend = true;                //是否显示数据说明
Chart1.Series[0].MarkerStyle = MarkerStyle.Circle;        //线条上的数据点标志类型
Chart1.Series[0].MarkerSize = 8;                          //标志大小
dr.Close();
conn.Close();
```

说明：使用 Chart1.DataBindTable(dr,"性别");绑定数据,数据源是 DataReader,因为数据源中只有"性别"和"人数"两列,因此在调用 Chart1.DataBindTable 方法时告诉了图表 X 轴的名称为"性别","人数"被自动地设置为 Y 轴的数据。

例 8-5 表 8-2 是 2011—2015 年淡水河谷和力拓两个矿山铁矿石的产量,用 Chart 制作这两家矿山年份与产量的柱状图,效果如图 8-12 所示。

表 8-2 2011—2015 年世界两个矿山的铁矿石产量　　　　　　　　单位:万 t

年份	淡水河谷	力拓
2015	33339	26304
2014	31921	23355
2013	29979	20896
2012	30904	19886
2011	32263	19176

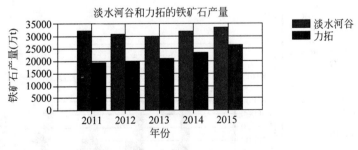

图 8-12 两个矿山铁矿石产量的柱状图

```
//清除所有绘图系列
Chart1.Series.Clear();
//增加标题
Title tt1 = new Title();
```

```
tt1.Text = "淡水河谷和力拓的铁矿石产量";
Chart1.Titles.Add(tt1);
//Series 默认的绘图区域 ChartAreas[0]
//增加两个绘图系列 Series1 和 Series2
Chart1.Series.Add("Series1");
Chart1.Series.Add("Series2");
//设置柱的颜色
Chart1.Series["Series1"].Color = Color.Red;
Chart1.Series["Series2"].Color = Color.Blue;
Chart1.Legends.Clear();                             //清除图例
Legend legendA = new Legend();                      //定义图例
Legend legendB = new Legend();
Chart1.Legends.Add(legendA);                        //将图例增加到 Chart 上
Chart1.Legends.Add(legendB);
Chart1.Series["Series1"].LegendText = "淡水河谷";    //设置图例的文字
Chart1.Series["Series2"].LegendText = "力拓";
//设置坐标轴
Axis AxisY = new Axis();
Axis AxisX = new Axis();
AxisX.Title = "年份";
AxisY.Title = "铁矿石产量(万 t)";
Chart1.ChartAreas[0].AxisX = AxisX;
Chart1.ChartAreas[0].AxisY = AxisY;
//Series1 的 Y、X 值
double[] yval1 = { 32263, 30904, 29979, 31921, 33339 };
string[] xval = { "2011", "2012", "2013","2014","2015" };
//Series2 的 Y 值
double[] yval2 = { 19176, 19886, 20896, 23355, 26304 };
//绑定数据
Chart1.Series["Series1"].Points.DataBindXY(xval, yval1);
Chart1.Series["Series2"].Points.DataBindXY(xval, yval2);
```

例 8-6 如果矿山铁矿石产量的数据是存储在 students 数据库的 mine 表中，数据如图 8-13(a)所示，用 Chart 制作矿山的年份与产量的柱状图，效果如图 8-13(b)所示。

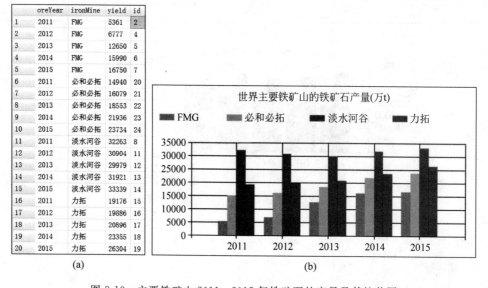

图 8-13 主要铁矿山 2011—2015 年铁矿石的产量及其柱状图

```
Title tt1 = new Title();
tt1.Text = "世界主要铁矿山的铁矿石产量(万 t)";
Chart1.Titles.Add(tt1);
string cnnString = WebConfigurationManager.ConnectionStrings["SQLServerString"].ConnectionString;
SqlConnection conn = new SqlConnection(cnnString);
string strSQL = "select * from mine";
SqlCommand cmd = new SqlCommand(strSQL, conn);
conn.Open();
SqlDataReader dr = cmd.ExecuteReader();
Chart1.DataBindCrossTable(dr, "ironMine", "oreYear", "yield", "LegendText = ironMine, Tooltip = yield ");
dr.Close();
conn.Close();
Chart1.Legends.Add("ironMine");                          //DataBindCrossTable 不自动添加图例
foreach (Legend aa in Chart1.Legends)
{
    aa.Docking = Docking.Top;                            //图例默认在右侧,改在图的上面
}
```

说明:

① DataBindCrossTable 有两个重载方法,分别如下:

```
public void DataBindCrossTable {
    IEnumerable dataSource,
    string seriesGroupByField,
    string xField,
    string yFields,
    string otherFields,
    PointSortOrder sortingOrder
}
public void DataBindCrossTable {
    IEnumerable dataSource,
    string seriesGroupByField,
    string xField,
    string yFields,
    string otherFields
}
```

各参数含义如下。

dataSource:要绑定的数据源。

seriesGroupByField:要分组统计的数据字段名称,如按姓名、日期等;示例中是 ironMine。

xField:X 轴绑定的字段名称。

yFields:Y 轴绑定的字段名称,如果需要绑定多个字段,则用逗号将字段名分开;示例中是 yield。

otherFields:其他要绑定的属性,使用的语法格式为

```
PointProperty = Field[{Format}] [,PointProperty = Field[{Format}]];
```

PointProperty 的取值可以是 AxisLabel、Tooltip、Label、LegendText、LegendTooltip 和 CustomPropertyName（自己定义的属性）。示例中"LegendText＝ironMine"用于设置图例上的文字，Tooltip 当移动到柱状图上时显示指定的字段，如示例中的 yield。

② DataBindCrossTable 方法不自动添加图例到 Chart，Chart1.Legends.Add("ironMine") 中"ironMine"是分组的字段名。

8.3.3 制作数据回归曲线

回归分析是数理统计的一个重要分支，在生产和科研中有着广泛的用途，如求经验公式、确定最佳生产条件、进行生产预测和控制等。回归分析根据所涉及的自变量的多少，可分为一元回归分析和多元回归分析；按照自变量和因变量之间的关系类型，可分为线性回归和非线性回归。

例 8-7 根据一元线性回归和非线性回归的基本方法，本书提供了以下 6 种回归方程的源程序 FittingFunct.cs。

$$y = a + bx \tag{1}$$

$$y = a + \frac{b}{\ln x} \tag{2}$$

$$y = a e^{\frac{b}{x}} \tag{3}$$

$$y = a e^{bx} \tag{4}$$

$$y = a x^b \tag{5}$$

$$y = \frac{x}{a + bx} \tag{6}$$

利用 FittingFunct.cs 程序，使用 Chart 控件，将 GradeRatio 表中的数据（图 8-14）以离散点的形式加以显示，绘制上述 6 种回归曲线并给出回归方程和相关系数，程序运行结果如图 8-15 所示。

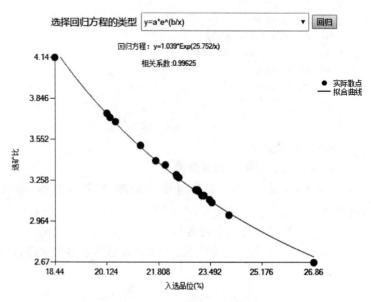

Grade	Ratio
23.10	3.18
22.46	3.28
20.41	3.68
23.07	3.19
23.01	3.19
21.71	3.40
23.21	3.15
20.13	3.74
21.22	3.51
22.40	3.29
20.23	3.71
18.44	4.14
22.38	3.30
23.28	3.15
26.86	2.67
23.46	3.12
22.02	3.37
23.54	3.10
24.10	3.01

图 8-14 GradeRatio 表中数据   图 8-15 使用 Chart 绘制回归曲线

❶ 新建 Web 窗体 Fitting.aspx，在窗体上拖放 Chart 控件 Chart1、DropDownList 控件 DropDownList1 和命令按钮 Button1，布局如图 8-16 所示。

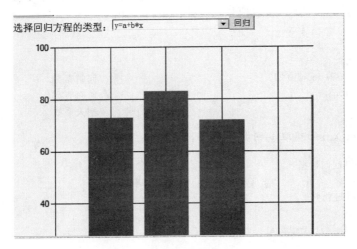

图 8-16　页面布局

❷ 设置 DropDownList1 属性 Items，如图 8-17 所示，图中"成员"的 Value 值分别为 1、2、3、4、5、6，Text 的值如图 8-17 所示。

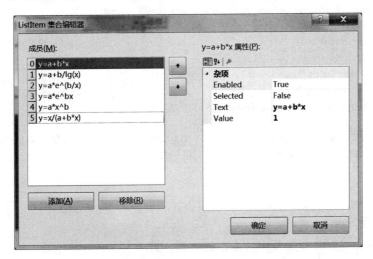

图 8-17　设置 DropDownList1 的 Items

❸ 引入回归程序的命名空间：

using asptest;　　//如果与当前程序是同一个命名空间，可省略

该命名空间中有一个类 FittingFunct，类中完成各种回归方程的静态函数有以下几个。

方程(1)　　Linear(double[] y,double[] x)：输出回归系数 a、b。

方程(2)　　LgEST(double[] y,double[] x)：输出回归系数 a、b。

方程(3)　　IndexEST(double[] y,double[] x)：输出回归系数 a、b。

方程(4)　　IndexEST1(double[] y,double[] x)：输出回归系数 a、b。

方程(5)　　PowerEST(double[] y,double[] x)：输出回归系数 a、b。

方程(6) HyperbolicEST(double[] y,double[] x)：输出回归系数 b、a。

计算相关系数 R：GetR(double[] y, double[] x, int equationNo)，输入参数是 y、x 数组，equationNo 是方程的编号，取值依次为 1、2、3、4、5、6。

❹ 定义 Web 窗体变量：

```csharp
double[] x,y;
double[] ab = new double[2];                    //回归系数
string equationStr = "";                        //回归方程
double R = 0;                                   //相关系数
```

❺ 定义 GetData() 获取要回归的数据：

```csharp
#region 获得回归数据
///<summary>
///获得要回归数据
///</summary>
private void GetFittingData()
{
    string cnnString = WebConfigurationManager.ConnectionStrings["SQLServerString"].ConnectionString;
    SqlConnection conn = new SqlConnection(cnnString);
    string strSQL = "select * from gradeRatio";
    DataTable dt = new DataTable();
    SqlDataAdapter adapter = new SqlDataAdapter(strSQL, conn);
    adapter.Fill(dt);
    if (dt.Rows.Count > 0)                      //有相关模型的数据
    {
        x = new double[dt.Rows.Count];
        y = new double[dt.Rows.Count];
        for (int i = 0; i < dt.Rows.Count; i++)
        {
            x[i] = Convert.ToDouble(dt.Rows[i][0].ToString());   //第1列是表中自变量
            y[i] = Convert.ToDouble(dt.Rows[i][1].ToString());   //第2列是表中因变量
        }
    }
}
#endregion
```

❻ 编写一个得到回归系数和相关系数的函数：

```csharp
#region 得到回归系数
private void GetAB(int EquationNo)
{
    //返回回归系数
    switch (EquationNo)
    {
        case 1:
            ab = FittingFunct.Linear(y, x);     //a = ab[0],b = ab[1]是回归方程的输出系数
            R = FittingFunct.GetR(y,x,1);
            break;
        case 2:
            ab = FittingFunct.LgEST(y,x);
```

```
                    R = FittingFunct.GetR(y, x, 2);
                    break;
                case 3:
                    ab = FittingFunct.IndexEST(y, x);
                    R = FittingFunct.GetR(y, x, 3);
                    break;
                case 4:
                    ab = FittingFunct.IndexEST1(y, x);
                    R = FittingFunct.GetR(y, x, 4);
                    break;
                case 5:
                    ab = FittingFunct.PowerEST(y, x);
                    R = FittingFunct.GetR(y, x, 5);
                    break;
                case 6:
                    ab = FittingFunct.HyperbolicEST(y,x);
                    R = FittingFunct.GetR(y, x, 6);
                    break;
        }
        //return ab;
    }
#endregion
```

❼ 根据自变量 x 和回归方程，计算因变量 y 的值，给出回归方程：

```
#region 根据自变量x,计算回归值
///<summary>
///得到回归值
///</summary>
///<param name = "EquationNo">EquationNo 回归方程编号</param>
///<param name = "varX">自变量 x </param>
///<returns></returns>
private double GetFittingValue(int EquationNo,double varX)
{
    double value = 0;
    switch (EquationNo)
    {
     case 1:
        value = ab[0] + ab[1] * varX;
        if (ab[1] > 0)
        {
            equationStr = "回归方程：y = " + ab[0].ToString("0.000") + " + " + ab[1].ToString("0.000") + " * x";
        }
        else
        {
            equationStr = "回归方程：y = " + ab[0].ToString("0.000") + " - " + Math.Abs(ab[1]).ToString("0.000") + " * x";
        }
        break;
     case 2:
```

```csharp
                    value = ab[0] + ab[1]/(Math.Log(varX));
                    if (ab[1] > 0)
                    {
                        equationStr = "回归方程: y = " + ab[0].ToString("0.000") + " + " + ab[1].ToString("0.000") + "/ ln(x)";
                    }
                    else
                    {
                        equationStr = "回归方程: y = " + ab[0].ToString("0.000") + " - " + Math.Abs(ab[1]).ToString("0.000") + "/ ln(x)";
                    }
                    break;
                case 3:
                    value = ab[0] * Math.Exp(ab[1]/varX);
                    if(ab[1]< 0)
                    {
                        equationStr = "回归方程: y = " + ab[0].ToString("0.000") + " * Exp(" + " - " + Math.Abs(ab[1]).ToString("0.000") + "/x)";
                    }
                    else
                    {
                        equationStr = "回归方程: y = " + ab[0].ToString("0.000") + " * Exp(" + Math.Abs(ab[1]).ToString("0.000") + "/x)" ;
                    }
                    break;
                case 4:
                    value = ab[0] * Math.Exp(ab[1] * varX);
                    equationStr = "回归方程: y = " + ab[0].ToString("0.000") + " e^(" + ab[1].ToString("0.000") + " * x)";
                    break;
                case 5:
                    value = ab[0] * Math.Pow(varX,ab[1]);
                    if(ab[1]>0)
                    {
                        equationStr = "回归方程: y = " + ab[0].ToString("0.000") + " * " + "x^" + ab[1].ToString("0.000");
                    }
                    else
                    {
                        equationStr = "回归方程: y = " + ab[0].ToString("0.000") + " * " + "x^(" + ab[1].ToString("0.000") + ")";
                    }
                    break;
                case 6:
                    value = varX/(ab[1] + ab[0] * varX);
                    if (ab[1] > 0)
                    {
                        equationStr = "回归方程: y = " + "x/(" + ab[0].ToString("0.000") + ab[1].ToString("0.000") + " * x)";
                    }
                    else
                    {
                        equationStr = "回归方程: y = " + "x/(" + ab[0].ToString("0.000") + " - " + Math.Abs(ab[1]).ToString("0.000") + " * x)";
```

```
            },
            break;
        }
        return value;
    }
    #endregion
```

❽ Button1 下的代码如下:

```
protected void Button1_Click(object sender, EventArgs e)
{
    //x、y是回归的数据,可在数据库的相关表中读取
    GetFittingData();                        //读取要回归的数据
    if (x == null || x.Length == 0)
    {
        Response.Write("<script>alert('没有可供回归的数据!')</script>");
        return;                              //空值
    }
    double maxX,minX,maxY,minY;              //x,y最大值,最小值
    maxX = x.Max();
    minX = x.Min();
    maxY = y.Max();
    minY = y.Min();
    Chart1.Series.Clear();
    Chart1.Legends.Clear();
    Axis AxisY = new Axis();
    Axis AxisX = new Axis();
    AxisY.Maximum = maxY;
    AxisY.Minimum = minY;
    AxisX.Maximum = maxX;
    AxisX.Minimum = minX;
    AxisX.Title = "入选品位(%)";
    AxisY.Title = "选矿比";
    AxisY.MajorGrid.Enabled = false;
    AxisY.IsStartedFromZero = false;
    AxisX.MajorGrid.Enabled = false;
    Chart1.Series.Add("seriesA");
    Chart1.Series.Add("seriesB");
    Chart1.ChartAreas[0].AxisX.MinorGrid.Enabled = false;
    Chart1.ChartAreas[0].AxisX.MajorGrid.Enabled = false;
    Chart1.ChartAreas[0].AxisX = AxisX;
    Chart1.ChartAreas[0].AxisY = AxisY;
    Legend legendA = new Legend("实际散点");
    Legend legendB = new Legend("拟合曲线");
    Chart1.Legends.Add(legendA);
    Chart1.Legends.Add(legendB);
    Chart1.Series["seriesA"].LegendText = "实际散点";
    Chart1.Series["seriesB"].LegendText = "拟合曲线";
    //指定序列 seriesA 的绘图类型是"点"
    Chart1.Series["seriesA"].ChartType = SeriesChartType.Point;
    //定义离散数据点的显示风格
```

```csharp
            MarkerStyle markerA = MarkerStyle.Circle;
            Chart1.Series["seriesA"].MarkerSize = 12;
            Chart1.Series["seriesA"].MarkerStyle = markerA;
            Chart1.Series["seriesA"].Color = Color.Blue;
            //指定序列 seriesB 的绘图类型是"线"
            Chart1.Series["seriesB"].ChartType = SeriesChartType.Spline;
            Chart1.Series["seriesB"].Color = Color.Red;
            int equationID = DropDownList1.SelectedIndex + 1;
            GetAB(equationID);                      //得到选择方程的系数,存放到数组 ab 中
            //x最大、最小值间分割成 100 份,通过这些点来画线
            double step = (maxX - minX) / 100;
            for (int i = 0; i < x.Length; i++)
            {
                Chart1.Series["seriesA"].Points.AddXY(x[i], y[i]); //画实际点
            }
            //画回归曲线
            for (double i = minX; i < maxX; i += step)
            {
                //画实际点
                Chart1.Series["seriesB"].Points.AddXY(i, GetFittingValue(equationID, i));
            }
            Title tt1 = new System.Web.UI.DataVisualization.Charting.Title();
            tt1.Text = equationStr;
            Title tt2 = new System.Web.UI.DataVisualization.Charting.Title();
            tt2.Text = "相关系数:" + R.ToString("0.00000");
            Chart1.Titles.Add(tt1);
            Chart1.Titles.Add(tt2);
            //保存图像为 aa.jpg
            Chart1.SaveImage(Server.MapPath("aa.jpg"), ChartImageFormat.Jpeg);
        }
```

8.4 TreeView 控件

ASP.NET 中用于页面导航的控件有 TreeView 控件、Menu 控件和 SiteMapPath 控件。本书只对 TreeView 控件做简单的介绍。

将 TreeView 控件拖放到页面,在页面"设计"视图中设置 TreeView 控件的 Nodes 属性,通过添加"根节点""子节点"的操作(图 8-18)生成树。

图 8-18 中增加节点后,在 VS 的"源"视图下,自动生成的以下标签:

```
<asp:TreeView ID="TreeView1" runat="server" Height="252px" Width="228px">
    <Nodes>
        <asp:TreeNode Text="武林高手" Value="武林高手">
            <asp:TreeNode Text="天龙八部" Value="天龙八部">
                <asp:TreeNode Text="乔峰" Value="乔峰"></asp:TreeNode>
                <asp:TreeNode Text="虚竹" Value="虚竹"></asp:TreeNode>
            </asp:TreeNode>
            <asp:TreeNode Text="射雕英雄传" Value="射雕英雄传">
                <asp:TreeNode Text="郭靖" Value="郭靖"></asp:TreeNode>
```

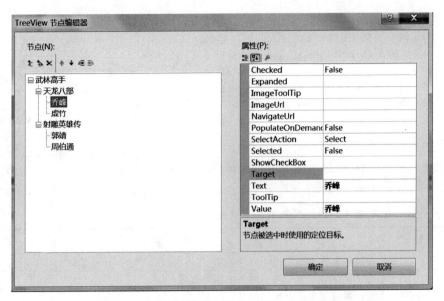

图 8-18 "设计"视图下为 TreeView 增加节点

```
            <asp:TreeNode Text = "周伯通" Value = "周伯通"></asp:TreeNode>
          </asp:TreeNode>
       </asp:TreeNode>
    </Nodes>
</asp:TreeView>
```

这种用法非常简单,也可以通过编程为 TreeView 增加上面的节点:

```
protected void Page_Load(object sender, EventArgs e)
{
    TreeNode rootNode = new TreeNode("武林高手");
    TreeView1.Nodes.Add(rootNode);              //增加根节点
    //定义两个子节点
    TreeNode childNodeA = new TreeNode();
    TreeNode childNodeB = new TreeNode();
    childNodeA.Text = "天龙八部";
    childNodeB.Text = "射雕英雄传";
    //在根节点下增加子节点
    rootNode.ChildNodes.Add(childNodeA);
    rootNode.ChildNodes.Add(childNodeB);
    //定义第 3 层节点
    TreeNode childNodeA1 = new TreeNode();
    TreeNode childNodeA2 = new TreeNode();
    TreeNode childNodeB1 = new TreeNode();
    TreeNode childNodeB2 = new TreeNode();
    childNodeA1.Text = "乔峰";
    childNodeA1.Value = "乔峰";
    childNodeA2.Text = "虚竹";
    childNodeA2.Value = "虚竹";
    childNodeB1.Text = "郭靖";
    childNodeB1.Value = "郭靖";
    childNodeB2.Text = "周伯通";
```

```
        childNodeB2.Value = "周伯通";
        //为两个子节点增加子节点
        childNodeA.ChildNodes.Add(childNodeA1);
        childNodeA.ChildNodes.Add(childNodeA2);
        childNodeB.ChildNodes.Add(childNodeB1);
        childNodeB.ChildNodes.Add(childNodeB2);
    }
```

图 8-19　为 TreeView 增加节点

运行上面的代码，结果如图 8-19 所示。

从上述代码可以看出，树上的每个节点是由 TreeNode 对象表示的。表 8-3 列出了 TreeNode 的主要属性。

表 8-3　TreeNode 的主要属性

属　性	说　明
Text	树节点上的文字
Value	用于保存节点上不显示出来的额外数据
ToolTip	光标停留在节点上提示的文字
NavigateUrl	用户单击该节点时自动转到该 URL；否则单击节点时会触发 TreeView.SelectedNodeChanged 事件
Target	在设置 NavigateUrl 的情况下，它会设置链接的目标窗口或者框架。不设置 Target，新页面在当前窗口打开
ImageUrl	显示在节点旁边的图片
ImageToolTip	显示在节点旁边图片的提示信息

例 8-8　根据 student、course、score 3 个表，将课程名 course_name 加入 TreeView 节点，展开课程名，子节点下显示出选修该课程的学生名（图 8-20），单击学生名，弹出该生的其他信息（图 8-21）。

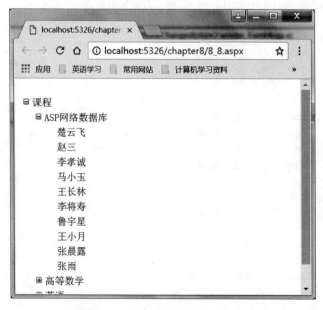

图 8-20　单击课程名打开子节点

图 8-21 单击最后节点弹出消息框

❶ 新建 Web 页面,在页面中拖放一个 TreeView 控件 TreeView1,在编程窗口引入命名空间:

```
using System.Data;
using System.Data.SqlClient;
using System.Web.Configuration;
```

❷ 定义以下窗体变量:

```
static string cnnString = WebConfigurationManager.ConnectionStrings["SQLServerString"].ConnectionString;
SqlConnection conn = new SqlConnection(cnnString);
SqlConnection conn = new SqlConnection(cnnString);
```

❸ 在 Page_Load 中写入如下代码:

```
if (!Page.IsPostBack)
{
    TreeNode rootNode = new TreeNode("课程");
    TreeView1.Nodes.Add(rootNode);                  //增加根节点
    string strSQLcourse = "select course_no,course_name from course";
    SqlDataAdapter adapterCourse = new SqlDataAdapter(strSQLcourse,conn);
    DataTable dtCourse = new DataTable();
    adapterCourse.Fill(dtCourse);
    for (int i = 0; i < dtCourse.Rows.Count;i++)
    {
        //根节点下增加"课程名"子节点
        TreeNode childNodeA = new TreeNode();
        childNodeA.Text = dtCourse.Rows[i]["course_name"].ToString();
        childNodeA.Value = dtCourse.Rows[i]["course_no"].ToString();
        rootNode.ChildNodes.Add(childNodeA);
```

```
            childNodeA.Expanded = false;              //不展开课程的子节点
            string strSQLStudent = "select score.student_no,score.course_no,
            student.student_name from score,course,student where
            score.course_no = course.course_no and score.student_no = student.student_no
            and score.course_no = '" + childNodeA.Value + "'";
            SqlDataAdapter adapterStudent = new SqlDataAdapter(strSQLStudent, conn);
            DataTable dtStudent = new DataTable();
            adapterStudent.Fill(dtStudent);
            for(int j = 0;j < dtStudent.Rows.Count;j++)
            {
                //为"课程名"增加子节点
                TreeNode childNodeAA = new TreeNode();
                childNodeAA.Text = dtStudent.Rows[j]["student_name"].ToString();
                childNodeAA.Value = dtStudent.Rows[j]["student_no"].ToString();
                childNodeA.ChildNodes.Add(childNodeAA);
            }
        }
    }
}
```

本程序在显示学生详细信息时,采用的是 Response.Write 输出 JavaScript 语句 alert() 的方式,它存在以下缺点:输出界面不美观,输出内容不能换行,输出时屏幕会出现空白。要解决此问题可采用 jQuery,详见第 9 章。

习题与思考

(1) 根据 student、course、score 3 个表,查询学生的成绩。程序运行后,在 DropDownList 框中自动增加 student_no,单击"查询"命令按钮,将该 student_no 的学生成绩以柱状图显示。柱状图 X 轴是该"学号"的学生所学的课程,Y 轴是课程的成绩,将光标移动到柱状图上,自动显示课程的成绩,如图 8-22 所示。

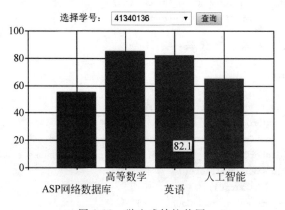

图 8-22 学生成绩柱状图

(2) 已知铁矿石的品位和密度(表 8-4),在 Chart 控件上绘制出密度与品位散点图。利用 FittingFunct.cs 程序,选用 6 种回归方程,以铁矿石的密度为自变量,品位为因变量,绘制回归曲线图,并给出相应的回归方程。

表 8-4　某矿山铁矿石的密度与品位数据

序号	密度/(g/cm³)	品位/%
1	3.45	25.1
2	3.48	27.3
3	3.54	29.4
4	3.62	31.6
5	3.76	35.2
6	3.78	38.2
7	3.88	41.6
8	3.96	45.3

（3）根据 teacher 和 course 数据表，使用 TreeView 控件，列出教师名，在教师名下增加子节点，列出该教师讲授的课程。

（4）使用 GridView 控件，根据 student、course、score 3 个表中内容，设计一个查询学生成绩的程序：选择学生的学号，查询出学生的学号、姓名、课程名、成绩。

（5）利用 UpLoad 控件，在 aspx 文件中编写将本地文件上传到自己所租云上的程序。

第 9 章

jQuery

(1) jQuery 基本语法。
(2) jQuery 事件。
(3) jQuery 动画效果。
(4) jQuery 操作表单和 HTML。
(5) jQuery 与 Ajax。
(6) jsTree、OrgChart、jquery.cookie 插件的使用。

在 Web 数据库程序开发时,经常为了实现一些美观的网页及震撼的异步刷新体验,需要 JavaScript 脚本编程的实现,但传统的 JavaScript 编程复杂、代码冗余,有时针对不同的浏览器,需要编写不同的脚本。jQuery 是一个专门用来动态改变 Web 页面文档的 JavaScript 库。使用 jQuery 只需少量的代码就可制作出复杂的效果,且开发出的程序支持各种主流浏览器。

9.1 jQuery 概述

9.1.1 jQuery 的作用

(1) 轻松获取 HTML 页面的元素。不用 jQuery,使用 JavaScript 遍历 DOM,查找 HTML 文档结构中某些特殊部分,要编写许多行代码,而利用 jQuery 提供的高效的选择符,可事半功倍。

(2) 动态修改页面外观和内容。CSS 虽然提供了修改页面外观的强大手段,但当浏览器不支持相同的标准时,单纯使用 CSS 显得力不从心,而 jQuery 提供了跨浏览器的标准解决方案;jQuery 用少量的代码,就可以改变文档的内容。

(3) 无刷新返回服务器端的数据信息。这种编程模式就是有名的 Ajax,jQuery 可以消除这一编程过程中根据不同浏览器定制不同程序代码的复杂性。

（4）增强页面的表现力。jQuery 可以轻松地给页面增加许多特技效果，如果使用 JavaScript 则需要大段的代码。

9.1.2 下载和引用 jQuery

1．下载

可以从 http://jquery.com/download/网站免费下载 jQuery 库。jQuery 有两种类型的不同版本，一种扩展名是.js 的，如 jquery-1.7.1.js，是未经过压缩的（可调试和阅读源代码）；另一种扩展名是.min.js 的，如 jquery-1.7.1.min.js，是压缩过的。书中使用的是 jquery.min.js，版本是 jQuery v1.11.3。

2．引用

在网页中要使用 jQuery 技术，需要将 js 文件引入到网页中。书中将 jQuery 库放在 Scripts 文件夹，Scripts 文件夹与各章节示例所在的文件夹是并列的，故在各章节网页的＜head＞＜/head＞之间，或者在＜body＞＜/body＞中，引用 jQuery 库（以引用 jquery.min.js 为例）：

```
<script type="text/javascript" src="../Scripts/jquery.min.js"></script>
```

9.1.3 用 jQuery 处理 DOM

用 JavaScript 处理 DOM 比较复杂，而 jQuery 可以让这种复杂变得简单。图 9-1 分别是用 JavaScript 与 jQuery 修改页面上第 1 个＜p＞中的内容，可以看出 jQuery 编写更加简单。$ 符号是 jQuery 函数的简单写法，即 jQuery("p")可简写为 $("p")。$()函数的主要工作就是获取括号中指定的元素。$()函数中可以放入 selector(选择符)，下面 3 种不同的内容增多可放在 $()中。

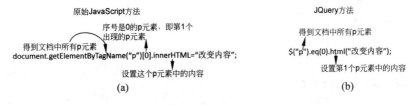

图 9-1　JavaScript 与 jQuery 语法上的不同

（1）CSS 选择器。如 $(".myClass")会返回 class 为 myClass 的元素集；$("#snoID")会返回 id="snoID"的所有元素。由于在同一个网页中，建议 id 不重复，故 $("#snoID")实际上返回的是某一个固定的 HTML 元素。

（2）HTML。如 $("img")会返回页面上全部的 img 元素。

（3）JavaScript 对象。如 $(document)返回当前文档、$(this)返回当前元素。

9.1.4 显示和隐藏小狗的示例

例 9-1　在浏览器中显示为图 9-2(a)，当单击"显示"按钮时，变为图 9-2(b)，再单击"隐藏"按钮,图9-2(b)图又变回到(a)图。

❶ 新建 HTML 网页 9_1.html：

```
1   <html>
2   <head><title>我家迪迪</title>
3   <link rel="stylesheet" type="text/css" href="9_1.css"/>
4   </head>
5   <body>
6   <div id="clickMe">显示</div>
7   <div id="picframe">
8      <img src="images/贵宾犬.jpg" alt="我家狗狗" />
9   </div>
10  <script type="text/javascript" src="../Scripts/jquery.min.js"></script>
11  <script>
12      $(document).ready(function () {
13          $("#clickMe").click(function () {
14              $("img").fadeIn(1000);
15              $("#picframe").slideToggle("slow");
16              if ($("#clickMe").html() == "隐藏")
17                  $("#clickMe").html("显示");
18              else
19                  $("#clickMe").html("隐藏");
20          });
21      });
22  </script>
23  </body>
24  </html>
```

图 9-2 显示和隐藏小狗

❷ 建立样式表 9_1.css：

```
#clickMe {
    background: #D8B36E;
    font-size:15pt;
    padding: 10px;
    text-align: center;
    width: 150px;
    display: block;
    border: 2px solid #000;
}

#picframe {
    background: #D8B36E;
    text-align: center;
    padding: 10px;
    width: 150px;
    display: none;
    border: 2px solid #000;
}
```

说明：

① 第 10 行,引用 jQuery 库文件 jquery.min.js。

② 第12行 $(document).ready(function (){}文档准备好,就执行function (){}中的代码。

③ 第13行 $("♯clickMe").click(function (){}),$("♯clickMe")是clickMe div 的id选择器,click(function (){})为这个元素的绑定事件 click。

④ 第14行 $("img").fadeIn(1000);,fadeIn(1000)是慢慢显示,时间是1000毫秒(1秒)。

⑤ 第15行 $("♯picframe").slideToggle("slow");,slideToggle("slow")上下切换函数,切换速度可设置为slow、fast、normal,或者毫秒。

⑥ 第16行 if ($("♯clickMe").html()=="隐藏")判断$("♯clickMe")的内容,JavaScript赋值运算符是"＝",比较运行符是"＝＝"。

9.2　jQuery选择器

1. $()

在jQuery中使用选择符可完成各种类型的jQuery操作,都要使用$(),()中存放的是各种类型的选择符,随后是对匹配的元素或元素集合应用jQuery提供的方法。如当光标滑到id为mydiv的元素上时,让该元素的文字变为红色;当光标移开时,该元素的文字变为黑色,可以使用以下代码:

```
$(document).ready(function(){
  $("♯mydiv").mouseover(function(){
      $(this).css("color","♯ff0000")
  });
  $("♯mydiv").mouseout(function(){
      $(this).css("color","♯000000")
  });
})
```

由于$()函数返回的是一个jQuery对象,故上面的代码可以使用串链,将各种方法链接到一起:

```
$(document).ready(function(){
  $("♯mydiv").mouseover(function(){
      $(this).css("color","♯ff0000")
  }).mouseout(function(){
      $(this).css("color","♯000000")
  });
})
```

再如将id="lighting"的元素,先执行fadeIn(),然后再执行fadeOut(),可以简写为:

```
$("♯lighting").fadeIn().fadeOut();
```

2. 元素选择器

网页元素由许多不同的HTML标签元素组成,jQuery可以直接选择网页元素以获取一个jQuery对象,进而可以使用jQuery提供的行为方法完成各种操作。

$("h1")选择网页中所有 h1 元素，$("h1").hide()中 hide()是方法，与对象间用"."连接，表示隐藏页面中所有 h1 元素，$("h1").css("color","♯FF0000")将网页中所有 h1 的颜色变为♯FF0000(红)色。对网页上 h1 元素可以这样操作，其他所有的 HTML 元素也都可以执行这样的操作。

3. 样式选择器

网页中元素样式的定义一般有两种方式，如<p class="title">，就可以使用$(".title")选择整个网页中全部 class="title"的元素，这种情况下想指定第 3 个 class="title"的元素，可以使用$(".title").eq(2)。如果为<p id="title">，由于网页中一般 id 不建议重复，故$("♯title")就是 id="title"的那个元素。

例 9-2 折叠式显示书的目录。浏览网页时，显示图 9-3 竖线左半部分，单击如"2 网页基础知识"链接后，显示竖线的右半部分。

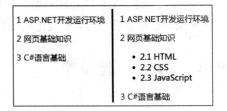

图 9-3 折叠式显示目录

新建网页 9_2.html，代码内容如下：

```html
<html>
<body>
<script type="text/javascript" src="../Scripts/jquery.min.js"></script>
<div>
    <p class="title">1 asp.net 开发运行环境</p>
    <ul class="subtitle">
        <li>1.1 Web 应用程序的工作原理</li>
        <li>1.2 静态网页和动态网页</li>
    </ul>
</div>
<div>
    <p class="title">2 网页基础知识</p>
    <ul class="subtitle">
        <li>2.1 HTML</li>
        <li>2.2 CSS</li>
        <li>2.3 JavaScript</li>
    </ul>
</div>
<div>
    <p class="title">3 C♯语言基础</p>
    <ul class="subtitle">
        <li>3.1 类的概念</li>
        <li>3.2 类的封装</li>
        <li>3.3 类的继承</li>
    </ul>
</div>
<script>
    $(document).ready(function () {
        $(".title").css("color", "♯ff0000");
        $(".subtitle").hide();
        $(".title").click(function () {
            $(this).next().toggle();
```

```
            })
        })
    </script>
</body>
</html>
```

说明:

① $(".title").css("color", "#ff0000");将所有的 class="title"的元素的颜色设置为红色。

② $(".subtitle").hide();隐藏所有 class="subtitle"的元素。

③ $(".title").click(function(){})为所有 class="title"的元素增加 click 事件。

④ $(this)指的是当前单击的 class="title"的元素,$(this).next()是当前单击元素的下一个元素,即当前单击的 class="title"元素下 ul 元素。

⑤ toggle()函数表示当前元素在显示与隐藏间切换:状态为显示时就变为隐藏;状态是隐藏时就变为显示。

4. 其他选择器举例

$(".title").children().eq(0):所有 class=".title"元素的后代的第 1 个元素。

$(".title").children().first():所有 class=".title"元素的后代的第 1 个元素。

$(".title").children().last():所有 class=".title"元素的后代的最后一个元素。

$(".title").parent().eq(0):所有 class=".title"元素的父代的第 1 个元素。

$(this).parent().parent():当前元素的父元素的父元素。

$(this).parent().next():当前元素的父元素的下一个元素。

$(this).prev():当前元素的上一个元素。

$("p.content"):选择所有 class="content"的<p>元素。

$(".title").find("li"):与 $(".title").children()相似,$(".title").children()是 class="title"的所有后代,而 $(".title").find()可指定 class="title"的某一级后代。

以下面的代码为例,说明 children()、eq()、first()、parent()、next()、prev()等函数的含义。借助这些函数,可以在 DOM 树上爬来爬去。

```
1   <body>
2   <div>
3       <p class="title">1 asp.net 开发运行环境</p>
4       <ul class="subtitle">
5           <li>1.1 Web 应用程序的工作原理</li>
6           <li>1.2 静态网页和动态网页</li>
7       </ul>
8   </div>
9   <div>
10      <p class="title">2 网页基础知识</p>
11      <ul class="subtitle">
12          <li>2.1 HTML</li>
13          <li>2.2 CSS</li>
14          <li>2.3 JavaScript</li>
15      </ul>
```

```
16  </div>
17  </body>
```

用法示例如下。

$(".subtitle").eq(0).children().eq(0)指的是第 5 行的<li>。

$(".subtitle").eq(1).parent()指的是第 9 行的<div>。

$(".subtitle").children().eq(0)指的是第 5 行的<li>。

$(".subtitle").children().eq(0).next()指的是第 6 行的<li>。

$(".subtitle").eq(0).prev()指的是第 3 行的<p>。

$("div").children()指的是第 3~7 行、第 10~15 行中的元素。

$("div").find("ul")指的是第 4~7 行、第 11~15 行中的元素。

一旦通过选择器确定了 HTML 中元素,要取得这些元素间的内容,可以使用 text()、html()方法。例如:

$(".subtitle").eq(0).text();

结果是:

1.1 Web 应用程序的工作原理
1.2 静态网页和动态网页

也可以使用 html()方法。例如:

$(".subtitle").eq(0).html();

结果是:

<li>1.1 Web 应用程序的工作原理</li>
<li>1.2 静态网页和动态网页</li>

9.3 jQuery 代码执行的时机和事件

jQuery 或者 JavaScript 编写的代码,触发运行的方式有两种:一种是系统自动触发,如将代码放在窗体的 load、unload 事件中,或者是将代码放在 $(document).ready(function(){})中;另一种是由用户键盘、鼠标等触发。

自动触发事件中的 window.onload 事件,当页面完全下载到浏览器(包括所有关联的文件)后触发,而 $(document).ready(function(){})在 DOM 上元素完全准备就绪后触发,触发时也许网页相关联的文件还没有下载完毕,故 $(document).ready(function(){})触发的时机要早于 window.onload 事件中的处理程序。

jQuery 中的事件与 JavaScript 中的事件相对应,只是写法上有些差异。JavaScript 中的事件以 on 开头,去除 on 后就变成 jQuery 中的事件。

jQuery 中的事件函数包括以下几个。

(1) 键盘事件:keydown、keyup、keypress。

(2) 鼠标事件:click、mousedown、mouseup、mouseover、mouseout、mouseenter、mouseleave、mousemove、dbclick、focusin、focusout、hover、toggle。

(3) 文档加载的事件: load、ready、unload。
(4) 表单事件: blur、change、focus、select、submit。
(5) 浏览器事件: error、resize、scroll。

以绑定 click 事件为例, jQuery 中绑定事件的方法有两种。

(1) $(selector).click(function(){代码;})。
(2) $(selector).bind('click',function(){代码;})。

如果 DOM 中的元素是通过代码创建的,如从数据库中读出学生的照片,将其显示在网页上,为每个照片绑定一个单击事件,单击后显示该学生的详细介绍,此时就只能使用方法(2)绑定事件。

9.4 jQuery 动态效果的函数

9.4.1 显示和隐藏

$(selector).hide(speed[,callback]): 隐藏元素。
$(selector).show(speed[,callback]): 显示元素。
$(selector).is(":hidden"): 判断元素是否隐藏。
$(selector).toggle(): 使用 show() 和 hide() 切换网页元素的显示和隐藏,隐藏显示的元素,显示隐藏的元素。

上面的参数中,speed 是显示或者隐藏速度,单位是毫秒(1 秒=1000 毫秒),速度还可表示为 slow、fast、normal。callback 为可选参数,表示 hide 和 show 执行完毕后被执行的函数名称,如 id="content" 的元素逐渐消失后出现对话框:

```
$(document).ready(function () {
    $("#content").hide(4000, function () {
        alert("我已经消失");
    })
})
```

如果写为:

```
$(document).ready(function () {
    $("#content").hide(4000);
    alert("我已经消失");
})
```

则会使 hide() 与 alert() 几乎同时发生,达不到先 id="content" 的元素逐渐消失,后出现对话框的效果。

9.4.2 滑动函数

$(selector).slideDown(speed[,callback]): 向下滑动显示元素。
$(selector).slideup(speed[,callback]): 向上滑动显示元素。
$(selector).slideToggle(speed[,callback]): 上下切换显示元素。

函数各参数的含义同 9.4.1 小节。

9.4.3 淡入淡出函数

$(selector).fadeIn(speed[,callback])：慢慢地显示。

$(selector).fadeOut(speed[,callback])：慢慢地隐藏。

$(selector).toggle(speed[,callback])：慢慢地隐藏显示的、显示隐藏的。

$(selector).fadeTo(speed,opacity[,callback])：设置元素的透明程度，参数 opacity 的取值范围为 0～1，透明程度由淡变清晰，对象由不清楚变清楚。

例 9-3 4 张图片开始时显示不清晰，当光标滑过图片时，图片变清晰，效果见图 9-4。

图 9-4 淡入淡出效果

新建 9_3.html 文件，代码内容如下：

```
<body>
    <script type="text/javascript" src="../Scripts/jquery.min.js"></script>
    <div>
    <img src="images/贵宾犬.jpg" /><img src="images/哈士奇.jpg"/>
    <img src="images/蝴蝶犬.jpg" /><img src="images/秋田犬.jpg"/>
    </div>
    <script>
        $(document).ready(function () {
            $("img").fadeTo("fast", 0.2);
            $("img").hover(function () { $(this).fadeTo("fast", 1); },
            function () { $(this).fadeTo("fast", 0.2); })
        })
    </script>
</body>
```

说明：hover()事件当光标滑进和滑出时触发，$(selector).hover(fnIn,fnOut)中 fnIn 表示光标滑进，fnIn 表示光标滑出。代码中 $("img").hover(function () { $(this).fadeTo("fast", 1); }, function () { $(this).fadeTo("fast", 0.2); })当光标滑进时图片最清楚，当光标滑出时图片变模糊。

9.5 jQuery 对 HTML/CSS 操作

使用 jQuery 从服务器端请求数据后，通过改变或者增加 HTML 内容的方式，将数据显示在浏览器上。

(1) 改变 HTML 内容。

$(selector).html(要改变的内容);

例如：

$("#mydiv").html("<p style='color:red'>新改变的内容</p>");

(2) 读取 HTML 内容。

var html=$(selector).html();

(3) 替换文本内容。

$(selector).text(要替换的内容);

(4) 读取文本内容。

var text=$(selector).text();

(5) 增加 HTML 内容。

$(selector).append(要增加的内容)：向所匹配的 HTML 元素的内部最后面增加内容。

$(selector).prepend(要增加的内容)：向所匹配的 HTML 元素的内部最前面增加内容。

$(selector).after(要增加的内容)：向所匹配的 HTML 元素之后增加内容。

$(selector).prepend(要增加的内容)：向所匹配的 HTML 元素之前增加内容。

(6) $(selector).insertAfter()。

在被选元素后插入 HTML 元素。例如：

$("button").click(function(){
  $("<span>Hello world!</span>").insertAfter("p");
});

会在每个<p>元素后插入一个<span>。

(7) $(selector).insertBefore()。

在被选元素前插入 HTML 元素。

(8) $(selector).attr()。

$(selector).attr(attribute)：返回属性的值。

$(selector).attr(attribute,value)：设置属性的值。

例如，设置图像的宽度：

$("img").attr("width","500");

再如，设置图像的 src：

$("img").attr("src","images\p1.jpg");

(9) $(selector).css()。

$(selector).css(property)：返回 CSS 属性值。

$(selector).css(property,value)：设置 CSS 属性和值。

例如，设置<p>的背景色为灰色：

$("p").css("background-color","grey");

9.6 jQuery 操作表单

9.6.1 表单中元素的选择器

$(":input")：选择所有输入元素。
$(":text")：选择所有类型为 text 的输入元素。
$(":radio")：选择所有类型为 radio 的输入元素。
$(":checkbox")：选择所有类型为 checkbox 的输入元素。
$(":submit")：选择所有类型为 submit 的输入元素。
$(":reset")：选择所有类型为 reset 的输入元素。
$(":checked")：选择所有未选中的输入元素。
$(":selected")：选择所有已经选择的输入元素。
$(":enabled")：选择所有启用的输入元素。
$(":disabled")：选择所有禁用的输入元素。
$(":password")：选择所有用于密码的输入元素。

9.6.2 jQuery 操作表单中的元素

jQuery 操作表单中的元素主要包括获取表单元素的值和为表单元素赋值两方面。下面根据表单元素，介绍这两方面的操作。

1. 操作文本框(text)

获取文本框的值：$("#text_id").val()。

为文本框赋值：$("#text_id").val("值")；或者 $("#text_id").attr("value","值")。

如果文本框只有 name 而没有 id，jQuery 通过以下方法获取文本值和赋值。

获取值：$("input[name='text_name']").val()。

赋值：$("input[name='text_name']").val("123")。

2. 操作单选按钮(radio)

以下面的 HTML 代码为例，加以说明：

```
<form id="ff">
<input type="radio" name="city" id="city1" value="1"/>
<label for "city1">北京</label>
<input type="radio" name="city" id="city2" value="2"/>
<label for "city2">天津</label>
<input type="radio" name="city" id="city3" value="3"/>
<label for "city3">上海</label>
```

```
< input type = "radio" name = "city" id = "city4" value = "4"/>
< label for "city3">重庆</label>
</form>
```

1) 通过属性 name 操作

① 获取选中 radio 的值：

```
var value = $("input:radio[name = 'city']:checked").val();
```

② 选中某个 radio：

```
$("input:radio[name = 'city']").eq(1).attr("checked", true);    //选中天津
```

或者

```
$("input:radio[name = 'city']").get(1).checked = true;           //选中天津
$("input:radio[name = 'city']:last").attr('checked', 'true');    //选中最后一个 radio
$("input:radio[name = 'city']:first").attr('checked', 'true');   //选中第一个 radio
```

③ 取消选中的某个 radio：

```
$("input:radio[name = 'city']").eq(1).attr("checked", false);    //不选天津
```

或者

```
$("input:radio[name = 'city']").get(1).checked = false
```

2) 通过属性 id 操作

① 获取选中 radio 的值：

```
$("#city1").val();
```

② 选中某个 radio：

```
$("#city1").attr("checked",true);
```

③ 不选中某个 radio：

```
$("#city1").attr("checked",false);
```

判断上述 radio 是否有选中项：

```
var chkRadio = $("input:radio[name = 'city']:checked").val();
if (chkRadio == null) {
alert("没有选中项");
    return false;
} else {
      alert(chkRadio);
}
```

3. 操作复选框(checkbox)

以下面的 HTML 代码为例：

```
< form id = "ff">
```

```html
< input type = "checkbox" name = "city" id = "city1" value = "1"/>
< label for "city1">北京</label >
< input type = "checkbox" name = "city" id = "city2" value = "2" checked = "checked"/>
< label for "city2">天津</label >
< input type = "checkbox" name = "city" id = "city3" value = "3"/>
< label for "city3">上海</label >
< input type = "checkbox" name = "city" id = "city4" value = "4"/>
< label for "city3">重庆</label >
</form >
```

① 选中第 3 个复选框：

```
$("input:checkbox[name = 'city']").get(2).checked = true;
```

或者

```
$("input:checkbox[name = 'city']").eq(2).attr("checked", true);
```

② 选中全部复选框：

```
$("input:checkbox[name = 'city']").each(function (index,value) {
    //方式 1
    //$("input:checkbox[name = 'city']").get(index).checked = true;
    //方式 2
    //$("input:checkbox[name = 'city']").eq(index).attr("checked", true);
    //方式 3
       $(this).attr("checked", true);
});
```

③ 复选框取消全部选中：

```
$("input:checkbox[name = 'city']").each(function (index) {
    //方式 1
    //$("input:checkbox[name = 'city']").get(index).checked = false;
    //方式 2
    //$("input:checkbox[name = 'city']").eq(index).attr("checked", false);
    //方式 3
    // $(this).attr("checked", false);
    //方式 4
       $(this).removeAttr("checked");
});
```

④ 获取复选框中选中项的 value 和 text：

```
var checkedValue = "";
var checkedText = "";
$("input:checkbox[name = 'city']:checked").each(function (index,value) {
    checkedValue += this.value + ",";
    checkedText += $(this).next().text() + ","
});
if( $("input:checkbox[name = 'city']:checked").length > 0)
{
    alert(checkedText.substr(0, checkedText.length - 1));
```

```
        alert(checkedText.substr(0, checkedValue.length - 1));
    }
```

也可以:

```
var checkedValue = "";
var checkedText = "";
$("input:checkbox[name = 'city']:checked").each(function (index) {
    checkedValue += $(this).val() + ",";
    checkedText += $(this).next().text() + ","
});
 if( $("input:checkbox[name = 'city']:checked").length > 0)
 {
    alert(checkedText.substr(0, checkedText.length - 1));
    alert(checkedText.substr(0, checkedValue.length - 1));
 }
```

4. 操作下拉列表框(select)

以下面的 HTML 代码为例,加以说明:

```
<form id = "ff">
    <select name = "city" id = "city">
        <option value = "1">北京</option>
        <option value = "2">天津</option>
        <option value = "3">上海</option>
        <option value = "4">重庆</option>
    </select>
</form>
```

① 设置下拉列表框中第 3 项为选中项:

```
$("#city")[0].selectedIndex = 2;
```

② 获取下拉列表框中选中项的 value 和 text:

```
var text = $("#city").find("option:selected").text();
var value = $("#city").find("option:selected").val();
```

或者

```
var value = $("#city option:selected").val();
var text = $("#city option:selected").text();
```

9.7 jQuery 与 Ajax

  Ajax 是一种与服务器交换数据的技术,可以在不重新装载整个页面的情况下更新网页的一部分。用 JavaScript 编写常规的 Ajax 代码并不容易,因为不同的浏览器对 Ajax 的实现并不相同。这意味着必须针对不同的浏览器编写额外的代码进行测试。借助 jQuery,只需要一行简单的代码就可以实现 Ajax 功能。下面结合实例介绍 jQuery 的部分 Ajax 方法的用法。

9.7.1 Ajax 方式提交数据

1. Ajax 提交表单数据前数据的格式化

使用 Ajax 方式向服务器发送信息之前,需要将表单上的数据加工处理成服务器上可以理解的数据格式,这样要把发送给服务器的数据串行化成为一个对象,以便 Ajax 将其作为一个数据包发送给服务器。jQuery 提供 serialize()和 serializeArray()两个方法完成串行化。下面以表单 myform 为例,说明这两个方法的区别。

```
<form id="myform">
<input type="text" name="a" value="1"/>
<input type="text" name="b" value="2"/>
<input type="hidden" name="c" value="3"/>
</form>
var sendData = $("#myform").serialize();
```

serialize()串行化会将所有表单的输入连接起来,构成一个键/值对组成的串,各个键/值对间用 & 号分隔,最终提交的 sendData 值是:

a=1$b=2&c=3

serializeArray()串行化将创建一个键/值对关联数组,其仍然是一个对象。$("#myform").serializeArray()后采用 alert(JSON.stringify(sendData))提交的 sendData 值是:

```
[
{"name":"a","value":"1"},
{"name":"b","value":"2"},
{"name":"c","value":"3"}
]
```

serialize()串行化方式简单,然而 serializeArray()数据结构更加清晰。

jQuery 串行化数据,要求表单上提交数据的控件设置 name 属性。表单上如果还有其他非 input 的控件,在串行化时可以:

```
$("#myform :input").serializeArray()
$("#myform :input").serialize()
```

$("#myform :input")中 myform 是表单 id 选择器,input 是 HTML 元素输入的过滤器,选择器只查找类型为 input 的 HTML 元素。

2. Ajax 向服务器发送数据

提交表单数据有 get 和 post 两种方式。对 ASP.NET 而言,客户端用 get 方法提交数据,服务器需要使用 Request.QueryString["a"]接收数据(在 ashx 文件中是 context.QueryString["a"]);客户端用 post 方法提交数据,则服务器需要用 Request.Form["a"]接收数据(在 ashx 文件中是 context.Form["a"]),由于二者易于混淆,ASP.NET 提供了 Request["a"]可接收 get 和 post 两种方式提交的数据。

jQuery 提供了以下几种向服务器发送和请求数据的方法,即$.get()、$.post()、$.getJSON()、$.Ajax(),前 3 个方法的参数和使用方法基本相同,只是$.getJSON 是专供 JSON 的一个快捷方法。

1) $.post()

基本语法格式为:

```
$.post(url_to_send,data,function(json){
   ...
})
```

各参数说明如下。

$：是 jQuery 的快捷方式。

post：表示提交数据的方式是 post。

url_to_send：表示数据要提交到这个 URL。

data：要提交的数据。

function()：提交完成后执行的回调函数。

2) $.get()

这个函数是对 get 方法提交数据的封装，只能使用在 get 提交数据解决异步刷新的方式上，使用方式和 $.post() 相同。

3) $.Ajax()

$.Ajax() 用法要比 $.get() 和 $.post() 复杂得多，下面列出的语法格式只包括它常用的参数。其语法格式为:

```
$.Ajax(url: '',
    method: 'get'|'post';
    cache:false|true;
    async:true|false;              //是否支持异步刷新,默认是 true
    data: {name:'yang',age:25},    //要提交的数据
    dataType: 'json'数据类型是 json
    beforeSend:function(xhr)
    { },
    success:function(data,textStatus,jqXHR)
    { },
    error(xhr,status,error)
    { },
    complete:function()
    { }
)
```

各参数说明如下。

method：提交给 URL 的方式，包括 get 或者 post。

url：通过 Ajax 获取数据的 URL。

cache：默认值为 true(dataType 为 'script' 或 'jsonp' 时,则默认为 false),指示是否缓存 URL 请求。如果设置为 false 将强制浏览器不缓存当前 URL 请求。该参数只对 HEAD、GET 请求有效(POST 请求本身就不会缓存)。

data：向 URL 请求数据时,提交给 URL 的数据。

dataType：向 URL 请求数据时,提交给 URL 的数据类型。

beforeSend(jpXHR,settings)：指定在请求发送前需要执行的回调函数。该函数还有两个参数：一是 jqXHR 对象，二是当前 settings 对象。这是一个 Ajax 事件，如果该函数返回 false，将取消本次 Ajax 请求。

从 jQuery 1.5 开始，$.ajax()方法返回 jQuery 自己的 XMLHttpRequest 对象(一般简称为 jqXHR)。

success()：指定请求成功后执行的回调函数。该函数有 3 个参数，即请求返回的数据、响应状态字符串、jqXHR 对象。

complete：指定请求完成(无论成功还是失败)后需要执行的回调函数。该函数还有两个参数：一个是 jqXHR 对象，一个是表示请求状态的字符串('success'、'notmodified'、'error'、'timeout'、'abort'或'parsererror')。

error：指定请求失败时执行的回调函数。该函数有 3 个参数，即 jqXHR 对象、请求状态字符串(null、'timeout'、'error'、'abort'和'parsererror')、错误信息字符串(响应状态的文本描述部分，如'Not Found'或'Internal Server Error')，这是一个 Ajax 事件。

对比$.post()与$.Ajax()就会发现，$.post()是简化了的$.Ajax()的数据提交方式，只能采用 POST 方式提交，只能是异步访问服务器，不能同步访问、不能进行错误处理、不能改变 cache 设置。使用$.post()时，$.Ajax()的主要几个参数，如 method、async 等进行了默认设置，是不可以改变的。

4) $.getJSON()

由于 JSON 非常流行且容易使用，jQuery 为其提供了一个特殊的快捷方法，即 getJSON()，专门用于 JSON 数据的获取，其语法格式为：

```
$.getJSON(url_to_load,data,function(json){
    ...
});
```

其中，url_to_load 是要加载数据的 URL；data 是传递给 URL 的数据，function(json){}是回调函数，json 是返回的数据，放在一个名为 json 的对象中。

对比$.getJSON()与$.ajax()可以看出，$.getJSON()实际上就是下面的$.ajax()：

```
$.ajax({
    url: url_to_load,
    dataType:'json',
    data:json,
    success:function(json){
        ...
    };
})
```

例 9-4 用$.getJSON()解析 introduce.json 文件，显示出小狗的编号、品种、介绍和照片，该文件的内容如下：

```
{
    "total":3,
    "rows":[
        {
```

```
            "petId":1,
            "petName":"贵宾犬",
            "headUrl":"href = '1.html'",
            "img":"src = 'images/贵宾犬.jpg'",
            "introduction":"活泼,性情优良,极易近人"
        },{
            "petId":2,
            "petName":"哈士奇",
            "headUrl":"href = '2.html'",
            "img":"src = 'images/哈士奇.jpg'",
            "introduction":"耐力和速度惊人,温顺友好"
        },{
            "petId":3,
            "petName":"蝴蝶犬",
            "headUrl":"href = '3.html'",
            "img":"src = 'images/蝴蝶犬.jpg'",
            "introduction":"活泼好动、胆大灵活"
        }
    ]
}
```

新建文件 9_4.html,内容如下:

```
<body>
    <script type = "text/javascript" src = "../Scripts/jquery.min.js"></script>
    <div id = "content"></div>
    <script>
        $(document).ready(function () {
            $.getJSON('introduce.json', function (data) {
                for (var i = 0; i < data.rows.length; i++) {
                    $('#content').append("<p>编号:" + data.rows[i].petId + ";品种:" + data.rows[i].petName + ";特点:" + data.rows[i].introduction + "</p>" + "<img" + data.rows[i].img + "' style = 'width:200px;height:200px'>");
                }
            })
        })
    </script>
</body>
```

上面程序虽然在 VS 中运行能够得到结果,但在 IIS 下却得不到结果。原因是 IIS 默认情况下不支持解析扩展名为 json 的文件。设置 IIS 支持 json 的方法(以 Windows 7 为例)如下。

进入 IIS,选中网站,选择"MIME 类型",单击"添加"按钮,弹出添加 MIME 类型对话框。在文件扩展名中输入.json,在 MIME 类型中输入 text/json。单击"确认"按钮,再重启网站即可。

3. 使用 load 方法

在 jQuery 中,load()方法通过 Ajax 请求从服务器加载数据,并把返回的数据放置在指定的元素中。使用 load()方法可以加载文本文件、网页文件、程序文件中的数据信息。

语法格式:

```
$(selector).load(url);
```

加载文本文件时,如果文本文件中有中文,要求将文本文件以 utf-8 的编码格式保存,而不是以默认的 ANSI 编码方式保存;否则 load()装入文本文件后会出现乱码。

例 9-5 新建 9_5.html 文件,运行后输入要查询的姓名,在当前页面显示查询结果,如图 9-5 所示。

图 9-5 查询运行结果

❶ 新建 9_5.html 文件,引用 jquery 库:

```
<script type="text/javascript" src="../Scripts/jquery.min.js"></script>
```

❷ 编写表单:

```
<div id="main">
    <form id="ff">
        请输入要查询的姓名<input type="text" name="sname"/>
    </form>
    <div id="result">
    </div>
    <button id="btnSearch">查询</button>
</div>
```

❸ 网页文档准备好后执行的 JavaScript 代码如下:

```
$(document).ready(function () {
    ("#btnSearch").click(function (event) {
        if ($("#sname").val().trim() == "" || $("#sname").val().trim() == null)
        {
            alert("必须填写查询的姓名");
            event.preventDefault();              //return false;
        }
        else
            btnSearch();
    })
})
```

说明:

① ("#btnSearch").click()为命令按钮绑定 click 事件时使用参数 event,通过 event.

preventDefault()取消对象默认的动作 click，保证必须输入查询姓名才能执行 click。当然如果使用 JavaScript 验证，return false；也能起到相同的效果。

② $("#sname").val()取得表单上文本框的值，trim()从字符串中移除前导空格、尾随空格和行终止符。

③ event.preventDefault()可以用 return false 取代。

❹ 查询 btnSearch()的代码如下：

```
function btnSearch()
{
    var sendData = $("#ff :input").serializeArray();
    $.post("9_5.ashx?m=" + Math.random(), sendData, function (returnData, status) {
        var jsonObj = eval('(' + returnData + ')');
        $("#result").html("");
        if (jsonObj.length > 0) {
            $.each(jsonObj, function (index, value) {
                $("#result").append("学号：" + value["student_no"] + "，姓名：" + value["student_name"] + "，身份证：" + value["identification"] + "<br/>");
            });
        }
        else {
            $("#result").html("查无此人");
        }
    });
}
```

说明：

① 提交表单数据时采用"9_5.ashx？m="＋Math.random()，在 URL 中加一个随机数传入，是为了防止页面缓冲而添加的，这是 Web 页面的通病，特别是程序在本地调试时，缓冲速度慢，代码修改了，可运行结果还是原来的样子，这时就可以通过这种方式改变URL。由于随机数每次都是不同的，这样每次都会和服务器交互一次而不是使用本地缓冲的数据库。

② JavaScript 的 eval()函数将服务器传来的 JSON 字符串转变为 JSON 对象，以便于解析 JSON；效果与 JSON.parse(returnData)相同，JavaScript 解析 JSON 参见第 3 章3.6 节。

③ 由于可能会有重名，故从数据库中查询到的结果可能不止一条记录，$.each(jsonObj,function(index,value)遍历集合 jsonObj，index 是索引值，value 是对应的值，读取 JSON 中对象的值时，要采用 value["JSON 对象中的属性名"]；当然也可采用例9_4.html 中的方法，读取 JSON 对象中的值。

❺ 后台服务器上 9_5.ashx 的代码如下：

```
public void ProcessRequest(HttpContext context)
{
    context.Response.ContentType = "text/plain";
    context.Response.Clear();
    string sname = context.Request.Form["sname"];
    if (DBbase.ConnectionSqlServer() == true)
```

```
            {
                List < studentData > students = new List < studentData >();
                students = DBbase.getStudentsByName(sname);
                JavaScriptSerializer js = new JavaScriptSerializer();
                string s = js.Serialize(students);
                context.Response.Write(s);
                DBbase.CloseSqlServer();
            }
        }
```

说明:

① 需要手工引入命名空间:

```
using System.Web.Script.Serialization;
using aspExample.DbEntity;                        //访问数据库的代码放在此命名空间中
```

② context.Request.Form["sname"]接收客户端传过来的 sname 值。

③ js.Serialize(students)将 students 序列化为字符串,参见第 5 章。

例 9-6 以读取 student 表中的记录为例,比较两种编程方式的不同:①不采用 Ajax 技术,直接将查询结果显示在 GridView 控件中;②将查询结果序列化为 JSON 字符串,并用 JavaScript 解析转化后的 JSON 字符串。

在.NET 中序列化和反序列化,可以使用 JSON.NET,其用法在第 5 章有详细的介绍。

由于在 DBbase.cs 中已经定义好了 getAllStudents()方法,该方法的返回值类型是 List<studentData>,故利用 JsonConvert.SerializeObject()方法,将 student 表中的记录序列化为 JSON 字符串。

❶ 下载 Newtonsoft.Json.dll 后,在 VS 的"解决方案资源管理器"中"引用"该文件。

❷ 新建 Web 窗体文件 9_6.aspx.cs,窗体上放置一个标签控件 Label1 和 GridView 控件 GridView1、命令按钮 Button1,设置命令按钮的 Text 为"服务器上请求数据"。

在 Page_Load 事件中写入以下代码:

```
protected void Page_Load(object sender, EventArgs e)
{
    Label1.Text = "现在时间是: " + DateTime.Now.ToString("yyyy - MM - dd hh:mm");
}
```

程序运行后在标签 Label1 上显示服务器当前的日期和时间。ToString("yyyy-MM-dd hh:mm")中 yyyy-MM-dd hh:mm 分别表示"年-月-日时:分"。

❸ 在命令按钮的 Click 事件中写入以下代码:

```
protected void Button1_Click(object sender, EventArgs e)
{
    if (DBbase.ConnectionSqlServer() == true)
    {
        List < studentData > students = new List < studentData >();
        students = DBbase.getAllStudents();
        DBbase.CloseSqlServer();
        GridView1.DataSource = students;
        GridView1.DataBind();
```

    }
}

DBbase.cs 中 ConnectionSqlServer()、getAllStudents()、CloseSqlServer()方法见第6章的6.5.2小节。getAllStudents()方法返回的是泛型集合,可以将其设置为 GridView1 的 DataSource。

上述②、③编程的方式,没有采用 Ajax 技术。程序运行后,单击命令按钮,结果见图 9-6。程序运行几分钟后,再次单击命令按钮,页面上的时间会发生变化,表明这种编程技术,当重新请求数据时,页面要全部刷新。

student_no	student_name	gender	email	identification	id
30012035	楚云飞	男	zsf@sohu.com	2303042006112015371	1
30012036	张雨	女	zy@sohu.com	2303042002100523462	2
40012030	赵三	男	zs@sohu.com	2303042000103023783	3
41321059	李孝诚	男	lxc@163.com	1101061998020300134	4
41340136	马小玉	女	mxy@126.com	1303041997051130285	5
41355062	王长林	男	wcl@163.com	1101061995120350126	6
41361045	李将寿	男	ljc@126.com	14510619901203503X	7
41361258	刘登山	男	lds@126.com	1300305200012342358	8
41361260	张文丽	女	zhangwl@163.com	1403052001062023489	9
41401007	鲁宇星	男	lyx@sohu.com	110108200012035012	10
41405002	王小月	女	wxy@263.com	130304199901115047	11
41405003	张晨露	女	zcl@263.com	110107199810038022	12
41405005	张伟达	男	zwd@263.net	120305199010270018	13
41405221	张康林	女	zkl@126.net	130305199805231248	14

现在时间是:2018-01-24 06:19 服务器上请求数据

图 9-6 ASP.NET 传统的控件编程模式

下面在同一个页面中,通过 jQuery 使用 Ajax,完成页面局部刷新,比较两种编程方式的不同。

❹ 在 9_6.aspx 的"源"视图中,在<head></head>中写入以下 HTML 代码:

< head runat = "server">
    < meta http - equiv = "Content - Type" content = "text/html; charset = utf - 8"/>
    < title></title>
    < link rel = "stylesheet" type = "text/css" href = "../Scripts/themes/default/easyui.css"/>
    < link rel = "stylesheet" type = "text/css" href = "..Scripts/themes/icon.css"/>
    < link rel = "stylesheet" href = "dist/themes/default/style.min.css" />
</head>

说明:本示例将使用 jQuery EasyUI 将 JSON 字符串直接显示在 easyui-datagrid 控件中。jQuery EasyUI 是在 jQuery 基础上提供的一种 Web 编程的技术框架,其网站 https://www.jeasyui.com/上有详细的使用文档和示例。本书第 12 章有 jQuery EasyUI 的使用说明。

❺ 在<body></body>中写入以下代码:

< script type = "text/javascript" src = "../Scripts/jquery.min.js"></script >
< script type = "text/javascript" src = "../Scripts/jquery.easyui.min.js"></script >

说明：jquery.easyui.min.js 是与 jquery.min.js 相匹配的 EasyUI 的 JavaScript 库。

❻ 将以下 HTML 代码加入到 `<body></body>` 中：

```html
<table id="dg" class="easyui-datagrid" style="width:600px;height:250px"
    title="学生记录" iconCls="icon-save"
    rownumbers="true" pagination="true">
    <thead>
    <tr>
        <th field="student_no" width="120">学号</th>
        <th field="student_name" width="120">姓名</th>
        <th field="email" width="150" align="right">email</th>
        <th field="identification" width="150" align="right">身份证编号</th>
    </tr>
    </thead>
</table>
    <button id="btnShow">easyUI 显示数据</button>
```

❼ 编写 jQuery 代码如下：

```html
<script>
    $(document).ready(function () {
        $("#btnShow").click(function () {
            getData();
        })
    })
    function getData()
    {
        $('#dg').datagrid({
            url: '9_6.ashx?m=' + Math.random()
        });
    }
</script>
```

说明：

① 在`<table>`中通过设置 class="easyui-datagrid"将表格转变为 datagrid，可以直接设置其 url "9_6.ashx"。"9_6.ashx"是一个输出的 JSON 字符串格式满足"easyui-datagrid"数据源的文件。本示例为了说明 Ajax 的局部刷新，是通过 JavaScript 客户端脚本指定 datagrid `<table>`的 URL。

② `<th field="student_no" width="120">学号</th>`中 field 属性的设置要与 "9_6.ashx"输出的 JSON 字符串中的属性相对应。

③ 以下脚本指定`<table>`的 URL：

```
$('#dg').datagrid({
    url: '9_6.ashx?m=' + Math.random()
});
```

❽ 在 9_6.ashx 文件中,引入以下命名空间:

```
using Newtonsoft.Json;
```

在 ProcessRequest()事件中输入以下代码:

```
public void ProcessRequest(HttpContext context)
{
    context.Response.ContentType = "text/plain";
    if (DBbase.ConnectionSqlServer() == true)
    {
        List < studentData > students = new List < studentData >();
        students = DBbase.getAllStudents();
        DBbase.CloseSqlServer();
        if (students.Count > 0)
        {
            string jsonStr = JsonConvert.SerializeObject(students);
            context.Response.Write("{\"rows\":" + jsonStr + "}");
            context.Response.End();
        }
    }
}
```

说明:

① string jsonStr = JsonConvert.SerializeObject(students);将 students 对象序列化为字符串,通过在语句 context.Response.Write("{\"rows\":"+jsonStr+"}");前设置断点,运行 9_6.aspx(注意:不是 9_6.ashx),监视到 jsonStr 的值如图 9-7 所示,可以看到输出的 JSON 字符串是数组格式的 JSON 字符串。

```
[{"student_no":"30012035","student_name":"楚云飞","gender":"男","email":"zsf@sohu.com","identification":"230304200611201537","id":1},
{"student_no":"30012036","student_name":"张雨","gender":"女","email":"zy@sohu.com","identification":"230304200210052346","id":2},
{"student_no":"40012030","student_name":"赵三","gender":"男","email":"zs@sohu.com","identification":"230304200010302378","id":3},
{"student_no":"41321059","student_name":"李孝诚","gender":"男","email":"lxc@163.com","identification":"110106199802030013","id":4},
{"student_no":"41340136","student_name":"马小玉","gender":"女","email":"mxy@126.com","identification":"130304199705113028","id":5},
{"student_no":"41355062","student_name":"王长林","gender":"男","email":"wcl@163.com","identification":"110106199512035012","id":6},
{"student_no":"41361045","student_name":"李将寿","gender":"男","email":"ljc@126.com","identification":"14510619901203503X","id":7},
{"student_no":"41361258","student_name":"刘登山","gender":"男","email":"lds@126.com","identification":"130030520001234235","id":8},
{"student_no":"41361260","student_name":"张文丽","gender":"女","email":"zhangwl@163.com","identification":"140305200106202348","id":9},
{"student_no":"41401007","student_name":"鲁宇星","gender":"男","email":"lyx@sohu.com","identification":"110108200012035012","id":10},
{"student_no":"41405002","student_name":"王小月","gender":"女","email":"wxy@263.com","identification":"130304199901115047","id":11},
{"student_no":"41405003","student_name":"张晨露","gender":"女","email":"zcl@263.com","identification":"110107199810038022","id":12},
{"student_no":"41405005","student_name":"张伟达","gender":"男","email":"zwd@263.net","identification":"120305199010270018","id":13},
{"student_no":"41405221","student_name":"张康林","gender":"女","email":"zkl@126.net","identification":"130305199805231248","id":14}]
```

图 9-7 表中记录 JSON 序列化后输出的内容

② 通过对 easyui-datagrid 的学习就会发现,easyui-datagrid 对 JSON 字符串的要求格式如下:

```
{"total":2, "rows":[{"student_no":"30012035","student_name":"楚云飞","gender":
"男","email":"zsf@sohu.com","identification":"230304200611201537","id":1},
{"student_no":"30012036","student_name":"张雨","gender":
"女","email":"zy@sohu.com","identification":"230304200210052346","id":2}
]}
```

将 easyui-datagrid 对 JSON 格式的要求与图 9-7 对比,需要在图 9-7 输出字符串前面拼接{"rows":(包括双引号),后面拼接},故:

```
context.Response.Write("{\"rows\":" + jsonStr + "}");
```

语句中\表示转义字符。

❾ 运行 9_6.aspx(不是 9_6.ashx),显示图 9-8。

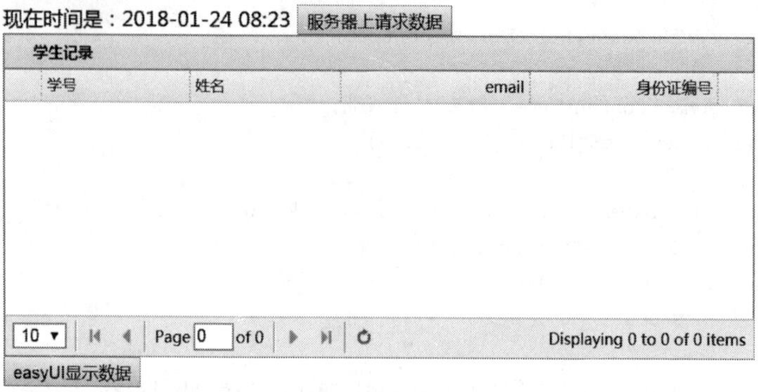

图 9-8 示例运行界面一

图 9-8 显示约 2 分钟后,单击"服务器上请求数据"按钮,显示图 9-9,注意到显示的时间发生了变化,表明页面重新进行了刷新。

在图 9-9 所示界面出现 2 分钟后,单击"easyUI 显示数据"按钮,显示图 9-10。图 9-10 中显示的时间与图 9-9 中显示的时间完全一致,表明使用 Ajax 技术可以保证页面局部刷新。

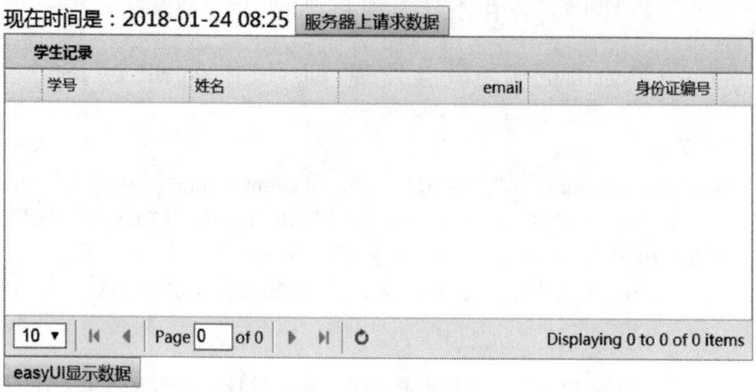

图 9-9 不使用 Ajax 技术从服务器请求数据

第9章　jQuery

student_no	student_name	gender	email	identification	id
30012035	楚云飞	男	zsf@sohu.com	230304200611201537	1
30012036	张雨	女	zy@sohu.com	230304200210052346	2
40012030	赵三	男	zs@sohu.com	230304200010302378	3
41321059	李孝诚	男	lxc@163.com	110106199802030013	4
41340136	马小玉	女	mxy@126.com	130304199705113028	5
41355062	王长林	男	wcl@163.com	110106199512035012	6
41361045	李将寿	男	ljc@126.com	14510619901203503X	7
41361258	刘登山	男	lds@126.com	130030520001234235	8
41361260	张文丽	女	zhangwl@163.com	140305200106202348	9
41401007	鲁宇星	男	lyx@sohu.com	110108200012035012	10
41405002	王小月	女	wxy@263.com	130304199901115047	11
41405003	张晨露	女	zcl@263.com	110107199810038022	12
41405005	张伟达	男	zwd@263.net	120305199010270018	13
41405221	张康林	女	zkl@126.net	130305199805231248	14

现在时间是：2018-01-24 08:25　服务器上请求数据

学生记录

	学号	姓名	email	身份证编号
1	30012035	楚云飞	zsf@sohu.com	230304200611201537
2	30012036	张雨	zy@sohu.com	230304200210052346
3	40012030	赵三	zs@sohu.com	230304200010302378
4	41321059	李孝诚	lxc@163.com	110106199802030013
5	41340136	马小玉	mxy@126.com	130304199705113028
6	41355062	王长林	wcl@163.com	110106199512035012
7	41361045	李孝寿	ljc@126.com	14510619901203503X

10 ▼　|◀　◀　Page 1 of 1　▶　▶|　⟳　　Displaying 1 to NaN of NaN items

easyUI显示数据

图 9-10　使用 Ajax 页面局部刷新示意图

图 9-10 中如果再次单击"服务器上请求数据"按钮，页面上的数据将全部重新刷新，页面回到图 9-9，只是显示的时间又发生了变化。

通过本示例，认识了页面的局部刷新，服务器传给客户端的 JSON 字符串，除了使用前面介绍的方法外，可以借助其他 JavaScript 类库，如 jQuery easyUI、jQuery UI 等，可以提供更好的用户界面。

例 9-7　网上提供许多免费的 JSON 数据接口，如天气预报、百度百科、电商各种服务的接口等，下面以查询城市天气预报为例，说明通过链接 http://www.sojson.com/open/api/weather/json.shtml?city=×××（×××表示城市名，该接口每天可免费调用 2000 次）提供的免费 API 接口，获取 JSON 数据并加以解析的过程。

请求某城市，如北京的天气成功后，返回的 JSON 格式如图 9-11 所示。

```
{"date":"20180130","message":"Success !","status":200,"city":"北京","count":1384,"data":
{"shidu":"13%","pm25":10.0,"pm10":25.0,"quality":"优","wendu":"3","ganmao":"各类人群可自由活
动","yesterday":{"date":"29日星期一","sunrise":"07:27","high":"高温 1.0℃","low":"低温
-8.0℃","sunset":"17:29","aqi":42.0,"fx":"西风","fl":"<3级","type":"晴","notice":"愿你拥有比阳光明媚的心
情"},"forecast":[{"date":"30日星期二","sunrise":"07:26","high":"高温 5.0℃","low":"低温
-6.0℃","sunset":"17:30","aqi":32.0,"fx":"西北风","fl":"<3级","type":"晴","notice":"愿你拥有比阳光明媚的
心情"},{"date":"31日星期三","sunrise":"07:25","high":"高温 2.0℃","low":"低温
-8.0℃","sunset":"17:31","aqi":43.0,"fx":"西南风","fl":"<3级","type":"晴","notice":"愿你拥有比阳光明媚的
心情"},{"date":"01日星期四","sunrise":"07:24","high":"高温 3.0℃","low":"低温
-7.0℃","sunset":"17:32","aqi":60.0,"fx":"西北风","fl":"<3级","type":"多云","notice":"阴晴之间，谨防紫外
线侵扰"},{"date":"02日星期五","sunrise":"07:23","high":"高温 -1.0℃","low":"低温
-10.0℃","sunset":"17:34","aqi":34.0,"fx":"西风","fl":"3-4级","type":"晴","notice":"愿你拥有比阳光明媚
的心情"},{"date":"03日星期六","sunrise":"07:23","high":"高温 -1.0℃","low":"低温
-10.0℃","sunset":"17:35","aqi":25.0,"fx":"西北风","fl":"<3级","type":"晴","notice":"愿你拥有比阳光明媚
的心情"}]}}
```

图 9-11　请求北京的天气预报返回的 JSON

返回的 JSON 如果不解析，阅读起来不方便。解析今天和明天的数据后，运行结果如图 9-12 所示。

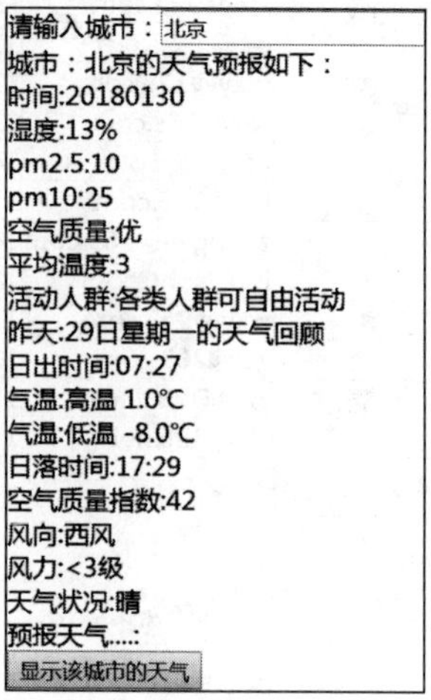

```
请输入城市：北京
城市：北京的天气预报如下：
时间:20180130
湿度:13%
pm2.5:10
pm10:25
空气质量:优
平均温度:3
活动人群:各类人群可自由活动
昨天:29日星期一的天气回顾
日出时间:07:27
气温:高温 1.0℃
气温:低温 -8.0℃
日落时间:17:29
空气质量指数:42
风向:西风
风力:<3级
天气状况:晴
预报天气……:
[显示该城市的天气]
```

图 9-12　解析部分 JSON 后的天气预报

如果在图 9-12 中输入其他城市，也可得到该城市的天气预报。

❶ 新建 9_7.html 文件，该文件的内容如下：

```html
<!DOCTYPE html>
<html xmlns="http://www.w3.org/1999/xhtml">
<head>
<meta http-equiv="Content-Type" content="text/html; charset=utf-8"/>
    <title></title>
</head>
<body>
    <script type="text/javascript" src="../Scripts/jquery.min.js"></script>
    <form id="ff">
        请输入城市：<input id="city" type="text" name="city" value="北京"/>
    </form>
    <div id="content"></div>
    <button onclick="showWeather()">显示该城市的天气</button>
    <script>
        function showWeather() {
            var data = $("#ff :input").serialize();
            $.post("9_7.ashx?m=" + Math.random(), data, function (returnData) {
                var jsonObj = eval('(' + returnData + ')');
                $("#content").html("");
                $("#content").append("城市：" + jsonObj.city + "的天气预报如下：<br>");
                $("#content").append("时间:" + jsonObj.date + "<br>");
                $("#content").append("湿度:" + jsonObj.data.shidu + "<br>");
                $("#content").append("pm2.5:" + jsonObj.data.pm25 + "<br>");
                $("#content").append("pm10:" + jsonObj.data.pm10 + "<br>");
                $("#content").append("空气质量:" + jsonObj.data.quality + "<br>");
                $("#content").append("平均温度:" + jsonObj.data.wendu + "<br>");
                $("#content").append("活动人群:" + jsonObj.data.ganmao + "<br>");
                $("#content").append("昨天:" + jsonObj.data.yesterday.date + "的天气回顾<br>");
```

```
            $("#content").append("日出时间:" + jsonObj.data.yesterday.sunrise + "<br>");
            $("#content").append("气温:" + jsonObj.data.yesterday.high + "<br>");
            $("#content").append("气温:" + jsonObj.data.yesterday.low + "<br>");
            $("#content").append("日落时间:" + jsonObj.data.yesterday.sunset + "<br>");
            $("#content").append("空气质量指数:" + jsonObj.data.yesterday.aqi + "<br>");
            $("#content").append("风向:" + jsonObj.data.yesterday.fx + "<br>");
            $("#content").append("风力:" + jsonObj.data.yesterday.fl + "<br>");
            $("#content").append("天气状况:" + jsonObj.data.yesterday.type + "<br>");
            $("#content").append("预报天气…:");
            return false;
        })
    }
</script>
</body>
</html>
```

❷ 新建9_7.ashx文件,需要增加以下命名空间:

```
using System.Net;
using System.IO;
using System.Text;
```

❸ 编写程序代码如下:

```
public void ProcessRequest(HttpContext context)
{
    context.Response.ContentType = "text/plain";
    string city = context.Request["city"].Trim();
    string strURL = "http://www.sojson.com/open/api/weather/json.shtml?city=" + city;
    var json = getHTML(strURL);
    context.Response.Write(json);
    context.Response.End();
}

string getHTML(string URL)
{
    string r = "";
    try
    {
        WebRequest wrGETURL = WebRequest.Create(URL);
        wrGETURL.Proxy = null;
        Stream objStream = wrGETURL.GetResponse().GetResponseStream();
        StreamReader objReader = new StreamReader(objStream, Encoding.GetEncoding("utf-8"));
        r = objReader.ReadToEnd();
    }
    catch (WebException e)
    {
        throw new Exception(e.Message);
    }
    catch (Exception e)
    {
        throw new Exception(e.Message);
```

```
        }
        return r;
}
```

说明：WebRequest 类在命名空间 System.Net 中是一个 abstract 类，用于对统一资源的标识符(URI)发出请求，其 Create()方法为指定的 URI 方案初始化新的 WebRequest 实例。GetResponse()返回对网络请求的响应，GetResponseStream()得到从服务器返回网页内容的数据流。

9.7.2 浏览器解析 XML 数据

服务器与客户端可通过 XML 和 JSON 进行数据传输。除了掌握 jQuery 如何解析 JSON 数据，也需要掌握 jQuery 如何利用 $.Ajax()在浏览器上解析 XML 数据的方法。

jQuery 中经常需要与一组元素逐个地交互。根据选择器选择的元素集合，可以通过循环遍历集合中的每一个元素，然后对每个元素分别处理。下面以例 9_2.html 中的 HTML 为例，说明 $(selector).each()的使用方法。

```
$(".title").each(function () {
    alert( $(this).text());
});
```

$(".title")得到所有 class=".title" 的 HTML 元素集，通过 each()遍历这个元素集合，遍历时执行函数 function(){}中的指令，$(this).text()指的是当前对象的文本值，故以上脚本执行后，对话框中的内容分别是：1 ASP.NET 开发运行环境、2 网页基础知识、3 C♯语言基础。在掌握 $(selector).each()的基础上，就可以使用 $.Ajax()解析 XML 了。

例 9-8 books.xml 内容如下：

```
<?xml version="1.0" encoding="utf-8"?>
<books>
    <book>
        <bname>Head First jQuery</bname>
        <author>Ryan Benedetti</author>
        <price>78.0</price>
    </book>
    <book>
        <bname>精通 C♯</bname>
        <author>Andrew Troelsen</author>
        <price>159.0</price>
    </book>
    <book>
        <bname>ASP.NET 基础教程</bname>
        <author>段克奇</author>
        <price>59.8</price>
    </book>
</books>
<head>
    <style>
```

```
            p{text-indent:20px;}
        </style>
    </head>
    <body>
    <script type="text/javascript" src="../Scripts/jquery.min.js"></script>
    <div id="tabs">
        <div id="computer">
            计算机书籍:
        </div>
        <div id="literature">
            文学书籍:
        </div>
    </div>
    <script>
        $.ajax({
            url: "books.xml",
            success: function (xml) {
                $(xml).find("book").each(function () {
                    if ($(this).find("classification").text() == "computer") {
                        var com = "<p>书名:" + $(this).find("bname").text() + $(this).find("bname").text() + ",作者:";
                        com += $(this).find("author").text() + ",单价:" + $(this).find("price").text() + "</p>"
                        $("#computer").append(com);
                    }
                    if ($(this).find("classification").text() == "literature") {
                        var lit = "<p>书名:" + $(this).find("bname").text() + ",作者:";
                        lit += $(this).find("author").text() + ",单价:" + $(this).find("price").text() + "</p>"
                        $("#literature").append(lit);
                    }
                })
            }
        });
    </script>
    /body>
    </html>
```

说明:

① $.Ajax()读取 books.xml 文件成功后,将结果放在回调函数的 xml 中,$(xml).find("book")是 books.xml 中所有 book 节点的集合,$(xml).find("book").each(function(){})遍历 book 节点。

② 在$(this).find("classification").text() == "computer"中,$(this)是当前的 book 节点。在当前 book 节点下找到 classification 节点的文本,判断其书籍的分类。如果书籍是"computer",就将书籍的相关信息添加到<div id="computer"></div>;如果是"literature",就将书籍的相关信息添加到<div id="literature"></div>。

解析后的结果如图 9-13 所示。

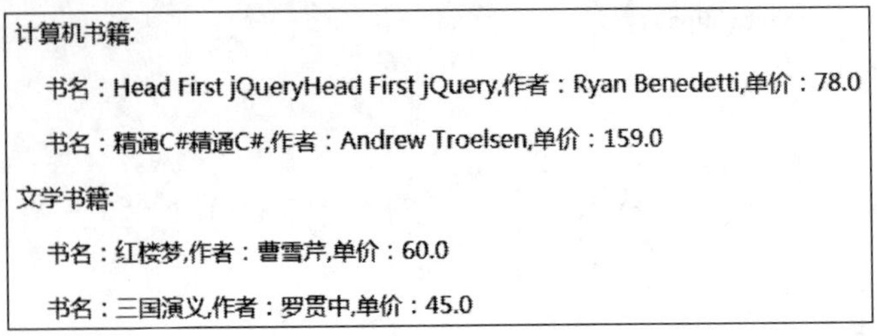

计算机书籍:
书名：Head First jQueryHead First jQuery,作者：Ryan Benedetti,单价：78.0
书名：精通C#精通C#,作者：Andrew Troelsen,单价：159.0
文学书籍:
书名：红楼梦,作者：曹雪芹,单价：60.0
书名：三国演义,作者：罗贯中,单价：45.0

图 9-13　解析 XML 结果

9.8　使用 jquery. cookie. js

采用 Ajax 方式编程，会遇到从一个 html 文件向另一个 html 文件传递变量的情况，如某些情况下，网页上的一些页面，只提供给特定的用户，没有授权的用户，通过 URL，直接输入这些网页的地址时，一般会给出“无使用权限”之类的提示。这种情况下可以考虑使用 cookies，jQuery 提供的 jquery. cookie. js 库可以方便地操作 cookies。

下载 jquery. cookie. js 的地址是 http://plugins. jquery. com/cookie/。本书中将下载后的文件放在 Script 的根文件夹下，使用时引用即可。

例 9-9　只有在图 9-14 的左半部分，输入用户名 zhang，输入密码 wang，提交后才能访问 9_10. html，如果用户名和密码不正确，或者直接访问 9_10. html，提示无权使用本网页，单击“确定”按钮后，将结束对本网页的访问，如图 9-14 的右半部分所示。

用户名:
密　码:
提交
重置
localhost:5326 显示：
您无权使用本页面！程序将退出
确定

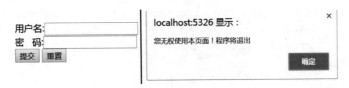

图 9-14　cookies 使用示例

❶ 新建 9_9. html 文件，假设是注册页面，网页中的内容如下：

```
<body>
<script type="text/javascript"src="../Scripts/jquery.min.js"></script>
<script type="text/javascript"src="../Scripts/jquery.cookie.js"></script>
<form>
    用户名:<input type="text" id="username" maxlength="15"/><br/>
    密   码:<input type="password" id="pwd" maxlength="15"/>
</form>
<button onclick="check()">提交</button>    <button>重置</button>
<script>
    function check()
    {
      if ($("#username").val()=="zhang" && $("#pwd").val()=="wang")
      {
        $.cookie('checkUser', 'yes', { expires: 2 });
        window.location="9_10.html";
```

```
        }
        else {
            $.cookie('checkUser', null);
            alert("用户名或密码错误!");
        }
    }
    </script>
</body>
```

说明:

① 在 $.cookie('checkUser', 'yes', {expires: 2}) 中, 将 yes 存储到名为 checkUser 的 cookies, 该 cookies 的有效期是 2 天。

② window.location="9_10.html" 转到页面 9_10.html。

❷ 新建受保护的页面如 9_10.html, 该页面中的内容如下:

```
<body>
<script type="text/javascript" src="../Scripts/jquery.min.js"></script>
<script type="text/javascript" src="../Scripts/jquery.cookie.js"></script>
<script language="javascript">
    if ($.cookie("checkUser") != "yes") {
        alert('您无权使用本页面!程序将退出');
        window.open('about:blank', '_self');
        window.close();
    }
</script>
<div>
    当你看到这些文字时, 表明你在 9_9 中输入的用户名是 zhang, 密码是 wang。
</div>
</body>
```

说明:

① <script type="text/javascript" src="../Scripts/jquery.cookie.js"></script> 引入 jQuery 的 cookies 库。

② 从 $.cookie("checkUser") 取出保存在 cookies 中名为 checkUser 的值。

③ 关闭窗口时, 没有直接用 window.close();, 而是增加语句 window.open('about:blank', '_self');, 用以避免关闭窗口时出现"是否关闭窗口"的对话框。

下面是 cookies 使用的介绍。

语法格式: $.cookie(名称, 值, [{option}])

(1) 读取 cookie 值:

$.cookie(cookieName)

其中, cookieName 为要读取的 cookie 名称。例如:

$.cookie("username"); 读取保存在 cookie 中名为的 username 的值

(2) 写入设置 cookie 值:

$.cookie(cookieName, cookieValue);

其中,cookieName:要设置的 cookie 名称;cookieValue:相对应的值。例如:

$.cookie("username","admin");将值"admin"写入 cookie 名为 username 的 cookie 中
$.cookie("username",NULL);销毁名称为 username 的 cookie

(3) [{option}]可选参数的说明。

path:cookie 值保存的路径,默认与创建页路径一致。

domin:cookie 域名属性,默认与创建页域名一样。这个地方要注意跨域的概念,如果要主域名和二级域名有效,则要设置".xxx.com"。

secrue:一个布尔值,表示传输 cookie 值时是否需要一个安全协议。

一个创建 cookie 的实例:

$.cookie("useuName", $("#useuName").val(),{path:"/", expiress: 7 ,sucue:true})

9.9 使用 jsTree 制作 tree

有时需要将数据以树型展示出来,除了使用 C♯ 提供给服务器端的 TreeView 控件外,也可在浏览器端使用 jQuery 的 jsTree 插件,其下载地址是 https://www.jstree.com/。本书中将其下载到 Scripts 文件夹下的 dist 文件夹中,dist 文件夹的内容如图 9-15 所示。

使用时在网页中引入该库文件:

< script src = "../Scripts/dist/jstree.min.js"></script >

或者

< script src = "../Scripts/dist/jstree.js"></script >

使用 jsTree 库时,还需要 jquery.min.js 库。

jsTree 插件的用法,在网址 https://www.jstree.com 上有各种 jsTree 应用的示例。

1. jsTree 基本使用方法

例 9-10 使用 jsTree,浏览器中显示如图 9-16 所示。

图 9-15 jsTree 所需的文件

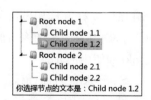

图 9-16 示例运行结果

❶ 新建 9_10.html 文件,将 jsTree 的样式文件链接到该页面。

```
< head >
< link rel = "stylesheet" href = "../Scripts/dist/themes/default/style.min.css"/>
</head >
```

❷ 建立可转化为 jsTree 的元素：

```html
<div id="jstree">
  <ul>
    <li>Root node 1
      <ul>
        <li id="c11">Child node 1.1</li>
        <li>Child node 1.2</li>
      </ul>
    </li>
    <li>Root node 2
      <ul>
        <li id="c21">Child node 2.1</li>
        <li>Child node 2.2</li>
      </ul>
    </li>
  </ul>
</div>
<div id="content"></div>
```

❸ 引用 jQuery 和 jsTree：

```html
<script type="text/javascript" src="../Scripts/jquery.min.js"></script>
<script src="../Scripts/dist/jstree.min.js"></script>
```

❹ 当 DOM 文档就绪后，建立 jsTree 实例，将❷中的 HTML 元素转化为 jsTree：

```javascript
$('#jstree').jstree();
```

❺ 绑定 jsTree 触发的事件，常用的是 changed.jstree，选择后触发该事件：

```javascript
$('#jstree').on("changed.jstree", function (e, data)
{
    $("#content").html("你选择节点的文本是：" + data.node.text);
}
);
```

此外，还有 $('#jstree').on("changed.jstree", function (e, data){}，在加载树时默认的某些操作执行，可以在这个函数中做很多操作。

❻ 事件中编写代码。如 $("#content").html("你选择节点的文本是：" + data.node.text);输出选中节点的文本。

示例中使用 JavaScript 编写的完整代码如下：

```javascript
<script>
    $(document).ready(function () {
        $('#jstree').jstree();
        $('#jstree').on("changed.jstree", function (e, data) {
            $("#content").html("你选择节点的文本是：" + data.node.text);
        });
    });
</script>
```

编程中,多数情况下树上的数据来源于数据库,是动态产生的,故在掌握 jsTree 基本使用方法后,需要学习如何将 JSON 数据绑定到 jsTree 上。

2. jsTree 使用 JSON 数据当作数据源

JSON 数据要在 jsTree 上正确地显示出来,其必须满足以下格式:

```
[
{"text":"root node","children": [{'text' : 'Child 1'},{'text' : 'Child 2'},{…}]},
{"text":"root node","children": [{'text' : 'Child 1'},{'text' : 'Child 2'},{…}]},
{…}
]
```

格式中"text"和"children"是其关键字,"root node"是"children"父节点上要显示的文本,'Child 1'、'Child 2'是子节点上要显示的文本。

例 9-11 使用 jsTree 将 course 表中的课程名以树的形式显示,如图 9-17(a)图所示,展开课程名,以树的形式显示出学习该课程的学生成绩,如图 9-17(b)图所示。

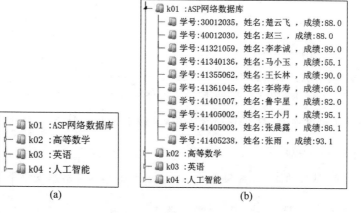

图 9-17 以 jsTree 显示课程名和该课程学生的成绩

❶ 新建 9_11.html 文件,该文件中内容如下:

```
<!DOCTYPE html>
<html xmlns = "http://www.w3.org/1999/xhtml">
<head>
<meta http-equiv = "Content-Type" content = "text/html; charset = utf-8"/>
<title></title>
<link rel = "stylesheet" href = "../Scripts/dist/themes/default/style.min.css"/>
</head>
<body>
  <script type = "text/javascript" src = "../Scripts/jquery.min.js"></script>
<script type = "text/javascript" src = "../Scripts/dist/jstree.min.js"></script>
  <div id = "TVCourse"></div>
    <script>
        $(document).ready(function () {
            $('#TVCourse').jstree({
                'core': {
                    'data': {
                        "url": "GetCourse.ashx?m = " + Math.random(),
```

```
                    "dataType": "json"
                }
            }
        });
    });
</script>
</body>
</html>
```

说明：

① <div id="TVCourse"></div>定义后,使用$('#TVCourse').jstree()将该<div>转化为jstree。

② "url": "GetCourse.ashx?m=" + Math.random()为jsTree指定数据源,为避免使用缓存数据,url字符串拼接了一个随机数。

❷ 建立一般处理程序 GetCourse.ashx,其代码如下：

```
public void ProcessRequest(HttpContext context)
{
    context.Response.ContentType = "text/plain";
    context.Response.Write(GetCourseName());
}
private string GetCourseName()
{
    if (DBbase.ConnectionSqlServer() == true)            //连接数据库
    {
    //需要拼出来以下格式:
    //[{"text":"root node","children":[{'text':'Child 1'},{'text' : 'Child 2'}]},...]
        //使用 StringBuilder 类完成字符串的拼接
        StringBuilder jsonBuilder = new StringBuilder();
        List<courseData> course = DBbase.getAllCourse();   //获取课程编号、课程名等
        jsonBuilder.Append("[");
        for (int i = 0; i < course.Count; i++)
        {
            jsonBuilder.Append("{");
            jsonBuilder.Append("\"text\":\"");                    //"text:"
            jsonBuilder.Append(course[i].course_no + ":" + course[i].course_name);
            jsonBuilder.Append("\",\"state\":{\"opened\":false}");   //"root node"
            DataTable dt = DBbase.getScoreByCourseNo(course[i].course_no);
            if(dt.Rows.Count == 0)
            {
                jsonBuilder.Append("}");
            }
            else
            {
                jsonBuilder.Append(",\"children\":[");
                //循环 children 下的节点
                for (int j = 0; j < dt.Rows.Count; j++)
                {
                    //拼第 2 层节点,格式
                    //"children": [{'text' : 'Child 1'},{'text' : 'Child 2'}]
```

```
                jsonBuilder.Append("{");
                jsonBuilder.Append("\"text\":\"");          //"text:"
                jsonBuilder.Append("学号:" + dt.Rows[j]["student_no"].ToString() + ",姓名:" +
    dt.Rows[j]["student_name"].ToString() + ",成绩:" + dt.Rows[j]["score"].ToString());
                jsonBuilder.Append("\"},");
            }
            //去除拼接串中最后的逗号
            jsonBuilder.Remove(jsonBuilder.Length - 1, 1);
            jsonBuilder.Append("],");
        }
        jsonBuilder.Remove(jsonBuilder.Length - 1, 1);
        jsonBuilder.Append("},");
    }
    jsonBuilder.Remove(jsonBuilder.Length - 1, 1);
    jsonBuilder.Append("]");
    DBbase.CloseSqlServer();                                //关闭数据库连接
    return jsonBuilder.ToString();
}
else
    return "";
}
```

说明:

① 使用 StringBuilder 类,需要引入命名空间 using System.Text。

② 通过调用 DBbase.cs 中 getAllCourse()方法(见 6.5.3 小节),将课程信息放到泛型集合 course 中。

③ 通过调用 DBbase.cs 中 getScoreByCourseNo()方法(见 6.5.3 小节),根据输入的 course_name,得到该课程所有学生的成绩。

④ 拼接字符串时,一定要按照 jsTree 对 JSON 格式的要求仔细拼写。

9.10 使用 OrgChart 制作组织结构图

编者在完成某边坡监测的科研项目时,需要将通信方式、监测设备单元、传感器等以图 9-18 所示的方式在网页中显示。

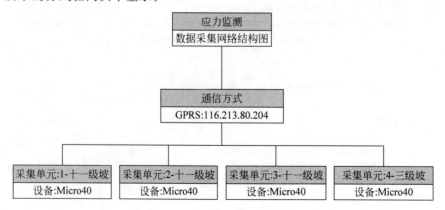

图 9-18 某边坡监测系统的显示

将光标移动到某采集单元,该采集单元根据当前节点是否具有子节点或同级节点,出现上、下、左、右箭头。单击向下箭头,出现图 9-19 中显示出该采集单元下布置的传感器及传感器目前运行的状态:灰色表示传感器已经启用,红色表示目前传感器停用。光标移动到某个传感器上,在该传感器右上角出现字符 i(如图中 04 传感器上),单击 i,显示出该传感器最新监测到的应力值、温度等参数。

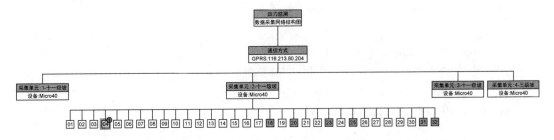

图 9-19　某边坡监测系统采集单元上布置的传感器

要在网页中实现以上功能,可以考虑 OrgChart。在官网 https://github.com/dabeng/OrgChart 上可以下载该插件及示例。通过学习 demo 下的示例,可快速地掌握其强大的功能。图 9-19 的示例可参考 demo 下的 option-createNode.html。

在本书的示例中,将下载的 jquery.orgchart.js 库放在 Script 文件夹下,css 文件夹、font 文件夹与 Script 文件夹平行。

例 9-12　利用 OrgChart 显示出教师的上课信息,如图 9-20 所示,当光标移动到有课程的教师节点上时,会出现向下箭头,单击该箭头,显示图 9-21。

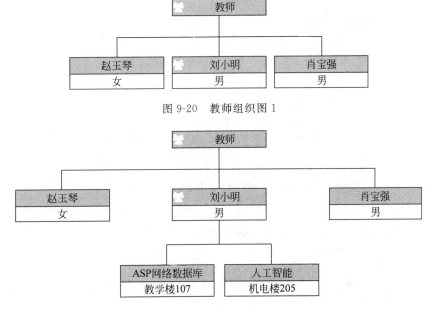

图 9-20　教师组织图 1

图 9-21　教师组织图 2

❶ 新建 HTML 页文件 9_12.html,在<head></head>中引入样式:

`<link rel="stylesheet" href="../css/font-awesome.min.css"/>`

```html
<link rel="stylesheet" href="../css/jquery.orgchart.css"/>
<link rel="stylesheet" href="../css/style.css"/>
```

引入样式时要注意当前文件和 css 文件夹的相对位置,由于当前文件是放在 chapter9 文件夹中,而 css 文件夹与 chapter9 是平行关系,故要使用..返回到上一层文件夹。

❷ 建立控制 OrgChart 显示格式的样式表,以控制 OrgChart 的背景色和类名为 teacherClass 的 content,使根节点只显示"教师"一行:

```html
<style>
    .orgchart {background: #fff;}
    .orgchart.teacherClass .content {display:none;}
</style>
```

❸ 在<body>中引入插件库:

```html
<script type="text/javascript" src="../Scripts/jquery.min.js"></script>
<script type="text/javascript" src="../Scripts/jquery.orgchart.js"></script>
```

❹ 在<body>中使用<div>,设定 OrgChart 要显示的位置:

```html
<div id="chart-container">
</div>
```

❺ 编写 JavaScript 代码:

```html
<script type="text/javascript">
    $(function() {
        var datasource = {
            'name': '教师',
            'title': '课程信息',
            'className':'teacherClass',
            'children': [
                {'name': '赵玉琴', 'title': '女'},
                {'name': '刘小明', 'title': '男',
                    'children': [
                        { 'name': 'ASP 网络数据库', 'title': '教学楼 107'},
                        { 'name': '人工智能', 'title': '机电楼 205'}
                    ]
                },
                {'name': '肖宝强', 'title': '男'}
            ]
        };
        $('#chart-container').orgchart({
            'data': datasource,
            'nodeContent': 'title',
            'visibleLevel': 2,
            'direction': 'r2l'
        })
    });
</script>
```

代码中使用 data 属性定义了 orgchart 对象的数据源是 datasource,nodeContent 属性

设置节点显示的文本内容，visibleLevel 设置展开的节点层数，direction 设置根节点的位置。

orgchart 对象的数据源是符合指定格式要求的 JSON 字符串，也可以是 JSON 对象。本示例中是 JSON 字符串。在多数情况下，需要从服务器的数据库中读取数据，将读取出的数据转变成 JSON，用 OrgChart 显示结果。

例 9-13　读取 students 库中 teacher 表和 course 表中的记录，显示图 9-22，单击的某教师节点上的向下箭头，显示出该教师的课程信息，如图 9-23 所示。

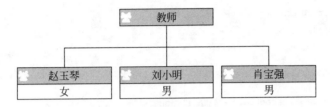

图 9-22　教师信息的组织图

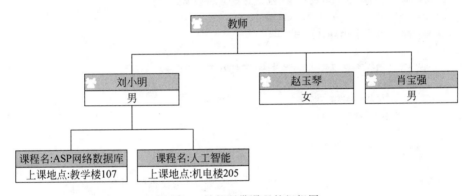

图 9-23　教师所带课程的组织图

❶ 新建 HTML 页文件 9_13.html，重复例 9-12 的前 4 个步骤（❶～❹），由于要显示的课程名可能较长，需要改变其默认的显示长度，在＜style＞中增加样式：

.orgchart .courseInfo{width:150px;}

❷ 在 9_13.html 中写入以下 JavaScript 代码：

```
<script type = "text/javascript">
    $(function () {
        $('#chart-container').orgchart({
            'data': "9_13.ashx?m = " + Math.random(),
            'nodeContent':'title',
            'visibleLevel': 2
        })
    });
</script>
```

指定 orgchart 插件的数据源来自一般处理程序 9_13.ashx，该程序将返回 JSON 格式。

❸ 在 6.5.1 小节中建立的 StudentDB.CS 文件中，新增加用于访问 teacher 表的 teacherData 类：

```csharp
public class teacherData
{
    public string teacher_ID
    {set; get;}
    public string teacher_name
    {set; get;}
    public string teacher_gender
    {set; get;}
    public string teacher_phone
    {set; get;}
}
```

❹ 在第6章中建立的 DBbase.cs 文件中,建立查询所有教师的方法 GetAllTeachers():

```csharp
public static List<teacherData> GetAllTeachers()
{
    List<teacherData> teachers = new List<teacherData>();
    string sql = "select * from teacher";
    if (ConnectionSqlServer() == true)
    {
        using (SqlCommand cmd = new SqlCommand(sql, conn))
        {
            SqlDataReader reader = cmd.ExecuteReader();
            while (reader.Read())
            {
                teachers.Add(new teacherData
                {
                    teacher_ID = reader["teacher_ID"].ToString(),
                    teacher_name = reader["teacher_name"].ToString(),
                    teacher_gender = reader["teacher_gender"].ToString(),
                    teacher_phone = reader["teacher_phone"].ToString(),
                });
            }
            reader.Close();
        }
        CloseSqlServer();
    }
    return teachers;
}
```

❺ 在第6章中建立的 DBbase.cs 文件中,建立根据教师编号查询课程信息的方法 getCourseByTeacherID():

```csharp
public static List<courseData> getCourseByTeacherID(string teacher_id)
{
    List<courseData> course = new List<courseData>();
    string sql = "select * from course where teacher_id = @teacher_id";
    SqlCommand cmd = new SqlCommand(sql, conn);
    cmd.Parameters.AddWithValue("@teacher_id", teacher_id);
    cmd.CommandText = sql;
    if (ConnectionSqlServer() == true)
```

```csharp
    {
        using (cmd)
        {
            SqlDataReader reader = cmd.ExecuteReader();
            while (reader.Read())
            {
                course.Add(new courseData
                {
                    course_no = reader["course_no"].ToString(),
                    course_name = reader["course_name"].ToString(),
                    course_address = reader["course_address"].ToString(),
                    course_time = reader["course_time"].ToString(),
                    teacher_ID = reader["teacher_ID"].ToString(),
                    id = (int)reader["id"]
                });
                        reader.Close();
            }
            CloseSqlServer();
        }
        return course;
    }
```

❻ 在 DBbase.cs 文件中建立为 OrgChart 输入指定格式的方法 JsonForOrgChart()：

```csharp
public static string JsonForOrgChart()
{
    StringBuilder jsonBuilder = new StringBuilder();
    List<teacherData> teacherData = GetAllTeachers();
    jsonBuilder.Append("{");
    jsonBuilder.Append("\"name\":\"教师\",\"title\":\"课程信息\",\"className\":\"teacherClass\"");
    if (teacherData.Count > 0)
    {
      jsonBuilder.Append(",\"children\":[");
      for (int i = 0; i < teacherData.Count; i++)
      {
         jsonBuilder.Append("{");
         jsonBuilder.Append("\"name\":\"");
         jsonBuilder.Append(teacherData[i].teacher_name);
         jsonBuilder.Append("\",\"title\":\"");
         jsonBuilder.Append(teacherData[i].teacher_gender);
         jsonBuilder.Append("\",\"className\":\"teacherInfo\"");
         List<courseData> courseData = getCourseByTeacherID(teacherData[i].teacher_ID);
         if (courseData.Count > 0)
         {
            jsonBuilder.Append(",\"children\":[");
            for (int j = 0; j < courseData.Count; j++)
            {
                jsonBuilder.Append("{");
                jsonBuilder.Append("\"name\":\"");
                jsonBuilder.Append("课程名:");
                jsonBuilder.Append(courseData[j].course_name);
                jsonBuilder.Append("\",\"title\":\"");
```

```
                jsonBuilder.Append("上课地点:");
                jsonBuilder.Append(courseData[j].course_address);
                jsonBuilder.Append("\",\"className\":\"courseInfo\"");
                jsonBuilder.Append("},");
            }
            jsonBuilder.Remove(jsonBuilder.Length - 1, 1);
            jsonBuilder.Append("]},");
        }
    }
    jsonBuilder.Remove(jsonBuilder.Length - 1, 1);
    jsonBuilder.Append("]}");
}
else
{
    jsonBuilder.Append("}");
}
return jsonBuilder.ToString();
}
```

使用 StringBuilder 的 jsonBuilder 对象拼接字符串时，需要拼接出例 9-12 中与 datasource 变量相类似的 JSON 格式。在输出 JSON 时节点中加入属性 className，以便控制各节点的显示格式。

❼ 新建在❷中指定数据源的一般处理程序 9_13.ashx，其代码如下：

```
public void ProcessRequest(HttpContext context)
{
    context.Response.ContentType = "text/plain";
    context.Response.Write(DBbase.JsonForOrgChart());
}
```

❽ 运行 9_13.html 文件。

上面的 OrgChart 示例，网页运行后自动加载组织结构图的所有节点，如果节点数量多，显示 OrgChart 时需要较长的等待时间。例 9-14 实现单击节点"赵玉琴"时为其增加子节点的功能，可参考 demo 文件夹下的 ondemand-loading-data.html。

例 9-14　网页运行后显示图 9-21，单击教师"赵玉琴"节点时，为其增加一个子节点，运行结果如图 9-24 所示。

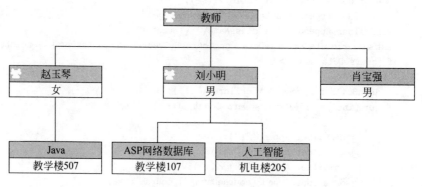

图 9-24　运行后增加子节点

❶ 新建 HTML 页文件 9_14.html，在<head></head>中引入以下样式：

```
< link rel = "stylesheet" href = "../css/font - awesome.min.css"/>
< link rel = "stylesheet" href = "../css/jquery.orgchart.css"/>
< link rel = "stylesheet" href = "../css/style.css"/>
```

❷ 建立控制 OrgChart 显示格式的样式表：

```
< style >
    .orgchart { background: #fff; }
    .orgchart.teacherClass.content { display:none;}
</style >
```

❸ 在<body>中引入插件库：

```
< script type = "text/javascript" src = "../Scripts/jquery.min.js"></script >
< script type = "text/javascript" src = "../Scripts/jquery.orgchart.js"></script >
< script type = "text/javascript" src = "../Scripts/jquery.mockjax.min.js"></script >
```

说明：jquery.mockjax.min.js 插件能够模拟 Ajax 请求，独立进行前台程序的开发，通过 https://www.npmjs.com/package/jquery-mockjax 下载插件，学习插件。

❹ 在<body>中使用<div>，设定 OrgChart 要显示的位置：

```
< div id = "chart - container">
</div >
```

❺ 编写 JavaScript 代码如下：

```
< script type = "text/javascript">
    $ (function () {
      $.mockjax({
        url: '/children/n2',
        contentType: 'application/json',
        responseTime: 1000,
        responseText:
        {
          'children':
          [
            { 'id': 'n21', 'name': 'Java', 'title': '教学楼 507', 'relationship': '110' }
          ]
        }
      });
      var datasource = {
        'id':'n1',
        'name': '教师',
        'title': '课程信息',
        'className': 'teacherClass',
        'relationship': '001',
        'children':
        [
          { 'id': 'n2', 'name': '赵玉琴', 'title': '女', 'relationship': '111' },
          {
            'id': 'n3', 'name': '刘小明', 'title': '男', 'relationship': '111',
            'children':
```

```
                    [
                        { 'id': 'n31', 'name': 'ASP 网络数据库', 'title': '教学楼 107', 'relationship': '110' },
                        { 'id': 'n32', 'name': '人工智能', 'title': '机电楼 205', 'relationship': '110' }
                    ]
                },
                { 'id': 'n6', 'name': '肖宝强', 'title': '男', 'relationship': '110' }
            ]
        };
        var ajaxURLs = {
            'children': '/children/',
            'parent': '/parent/',
            'siblings': '/siblings/',
            'families': '/families/'
        }
        $('#chart-container').orgchart({
            'data': datasource,
            'ajaxURL': ajaxURLs,
            'nodeContent': 'title',
            'nodeId': 'id',
            'visibleLevel': 3
        })
    });
</script>
```

说明：

① OrgChart 插件的 ajaxURL 属性具有 parent、children、siblings、families 4 个属性，为节点提供 Ajax 请求时的 URL。示例中指定为变量 ajaxURLs，属性 'children'：'/children/' 表示当节点的 URL 模板是 '/children/' 时，执行 mockjax 中能够与该模式的 URL 相匹配的 URL，为该节点增加子节点。示例中没有为 mockjax 设置 parent、siblings、families 的 URL 值。

② OrgChart 的 nodeId 属性，指定识别节点唯一性的属性。

③ 节点 'relationship' 表示当前节点与父节点、同级节点、子节点的关系，其值是一个用 0 和 1 表示的 3 位数。第 1 位用 1 表示该节点有父节点，0 表示没有父节点；第 2 位表示有无同级节点，第 3 位表示有无子节点，第 2 位和第 3 位值的含义同第 1 位。

示例中需要为 'id' 和 'n2' 的节点增加子节点，而该节点有父节点和同级节点，故设置其 'relationship' 值为 '111'。

④ $.mockjax 的 URL 属性指定为 '/children/n2'，URL 能够匹配 ajaxURL 的 'children' 属性的值 '/children/'，n2 是节点 id 的值。

⑤ $.mockjax 请求 URL 返回的值用 responseText 属性设置。

习题与思考

（1）有 4 张小狗的照片放在文件夹 chapter9/images 中，以循环的方式，将其显示在一个 HTML 的 image 元素中，每显示出一张照片，自动标出小狗的名称，如图 9-25 所示。

（2）在记事本中输入一段文字，然后保存为 mytxt.txt，使用 Ajax 将其读入到网页的指定位置，如<div id="content"></div>中。

这是贵宾犬　　　这是哈士奇　　　这是蝴蝶犬　　　这是秋田犬

图 9-25　循环依次显示狗狗的照片

（3）解析 JSON 中的数据，将全部"工号""姓名"信息填写到 <div id="content"></div> 中，结果见图 9-26，阅读程序，并填空。

```
工号：9001，姓名：张三
工号：8001，姓名：李四
工号：9002，姓名：王五
工号：7001，姓名：赵六
```

图 9-26　解析 JSON 后显示的结果

```
<!DOCTYPE html>
<html xmlns="http://www.w3.org/1999/xhtml">
<body>
    <script type="text/javascript" src="../Scripts/jquery.min.js"></script>
    <div id="content">
    </div>
    <script>
        var json = {
            "employees": [
                { "工号": "9001", "姓名": "张三", "部门": "人力部" },
                { "工号": "8001", "姓名": "李四", "部门": "财务部" },
                { "工号": "9002", "姓名": "王五", "部门": "账务部" },
                { "工号": "7001", "姓名": "赵六", "部门": "技术开发部" },
            ]
        };
        $("#content").empty();
        for(var i = 0; i < _____ ; i++)
        {
            $("#content").append(_____ + "<br>");
        }
    </script>
</body>
```

（4）以例 9-2.html 中的 HTML 为例编写脚本，获得网页上以下的内容，将其增加到 <div id="content"></div> 中，运行结果见下面的文字。

1.1　Web 应用程序的工作原理
1.2　静态网页和动态网页
2.1　HTML
2.2　CSS
2.3　JavaScript
3.1　类的概念概念
3.2　类的封装
3.3　类的继承

(5) 读下面的 HTML,根据要求,应用 jQuery 填写脚本:

```html
<!DOCTYPE html>
<html xmlns="http://www.w3.org/1999/xhtml">
<head>
<meta http-equiv="Content-Type" content="text/html; charset=utf-8"/>
    <title></title>
    <style>
        .fruit,.vegetable
        {
            display:block;float:left;
        }
        .myfruit
        {
            clear:both;
        }
    </style>
</head>
<body>
    <script type="text/javascript" src="../Scripts/jquery.min.js"></script>
    <div class="fruit">
       <ul>
           <li>lemon</li>
           <li>pear</li>
           <li>apple</li>
           <li>banana</li>
           <li>grape</li>
       </ul>
    </div>
    <div class="vegetable">
       <ul>
           <li>cabbage</li>
           <li>onion</li>
           <li>celery</li>
           <li>tomato</li>
           <li>cucumber</li>
       </ul>
    </div>
    <div class="myfruit" id="result"></div>
</body>
</html>
```

① 用脚本将菜单上的 onion 换成 lettuce：

_____

② 用脚本将菜单上 lemon 的背景色改为黄色：

_____

③ 用 $(selector).each() 将 class="fruit" 的所有 fruit 的文本读出来,写入 <div class="myfruit" id="result"></div> 中：

```
$(".fruit").find("li").each(function () {
```

_____;
})

（6）应用 jQuery 可以方便地控制表单上各元素，根据要求填空：

<body>
<script type="text/javascript" src="../Scripts/jquery.min.js"></script>
<form>
　　<input type="checkbox" name="vehicle" value="Ford" id="vehicle1"/>
<label for="vehicle1">Ford</label>
　　<input type="checkbox" name="vehicle" value="Cadillac" id="vehicle2"/>
<label for="vehicle2">Cadillac</label>
　　<input type="checkbox" name="vehicle" value="Hummer" id="vehicle3"/>
<label for="vehicle3">Hummer</label>
　　<input type="checkbox" name="vehicle" value="Dodge" id="vehicle4"/>
<label for="vehicle4">Dodge</label>
</form>
</body>

① 选中全部复选框的脚本：

$(_____).each(function () {
　　_____;
});

② 获得选中复选框的值，放到变量 checkedValue 的脚本：

var checkedValue = "";
　$(_____).each(function () {
　　checkedValue += _____ + ",";
});

（7）自学下载 OrgChart 的 demo 下的 vertical-level.html 文件和网站 https://github.com/dabeng/OrgChart，尝试着将图 9-23 所示的教师所带课程的组织图变为图 9-27。

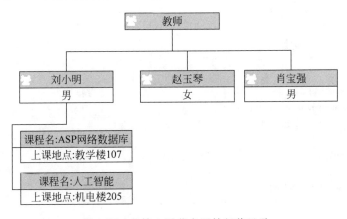

图 9-27　从第 3 层节点开始竖着显示

第 10 章

语言集成查询——LINQ

（1）LINQ to Objects 查询内存中的集合和数组。
（2）LINQ to XML 查询 XML 中数据。
（3）LINQ to Entities 操作数据库。

如果要从国家名 Vietnam、Japan、American、Korea、India、Australia、France 等中选出单词长度超过 5 个字母的国家名，且将检索结果排序。学习本章前，可以使用循环、判断等语句完成，代码可以如下：

```
string[] words = { "Vietnam", "Japan", "American", "Korea", "India", "Australia", "France" };
string[] result = new string[words.Length];
for (int i = 0; i < words.Length; i++)
{
    if (words[i].Length > 5)
    {
        result[i] = words[i];
    }
}
Array.Sort(result);                                    //排序
StringBuilder aa = new StringBuilder();
foreach (string bb in result)
{
    if (bb != null)
    {
        aa.Append(bb + "</br>");
    }
}
Response.Write(aa);
```

微软的 LINQ 提供了一种更加快捷的方式，可以从数据源中提取出数据子集。如果使用 LINQ，代码如下：

```
string[] words = { "Vietnam", "Japan", "American", "Korea", "India", "Australia", "France" };
var subwords = from i in words where i.Length > 5 orderby i select i;
StringBuilder aa = new StringBuilder();
foreach (string bb in subwords)
{
    aa.Append(bb + "</br>");
}
Response.Write(aa);
```

从代码上看,LINQ使我们绕开循环和判断,检索更加方便和容易了。需要说明的是,LINQ格式虽然与SQL的select语句有些像,但其与select语句完全不相干。

LINQ(Language Integrated Query)是从.NET Framework 3.5开始引入的一组功能,其提供了标准的、易于学习的查询和更新数据模式。程序员可以像SQL那样操作各种类型的数据库、XML文档及任何实现IEnumerable接口或者IEnumerable＜T＞泛型接口的.NET Framework集合类。

LINQ提供程序将LINQ查询映射到要查询的数据源。在编写LINQ查询时,提供程序接受该查询,并将其转换为数据源能够执行的命令。提供程序还将数据源中的数据转换为组合查询结果的对象。当向数据源发送更新命令时,能够将对象转换为数据。C♯包含以下LINQ提供程序。

(1) LINQ to Objects:提供程序可以查询内存中的集合和数组。

(2) LINQ to XML:提供查询和修改XML,既可以修改内存中的XML,也可以从文件中加载XML以及将XML保存为文件。

(3) LINQ to DataSet:提供查询和更新DataSet的功能。

(4) LINQ to SQL:提供查询和修改SQL Server等数据库的功能。

(5) LINQ to Entities:是Entity Framework的一部分,并且取代LINQ to SQL作为在数据库上使用的标准机制,它可以使用Entity Framework数据模型执行LINQ查询。

10.1 LINQ to Objects

10.1.1 LINQ查询语法和步骤

LINQ查询通常称为"查询表达式",由标识查询数据源的查询子句和标识查询迭代变量的查询子句组合而成。下面以一组数据为例说明LINQ查询的用法。

例10-1 一组数据5、12、8、20、25、15、18、16、13,将这组数据中偶数列出来,并且由小到大排序:

```
int[] numbers = { 5, 12, 8, 20, 25, 15, 18, 16, 13 };
var subset = from i in numbers where i % 2 == 0 orderby i select i;
foreach (int bb in subset)
{
    Response.Write("偶数有: " + "</br>" + bb);
}
```

说明：LINQ 查询分 3 个步骤，即获取数据源、创建查询、执行查询。

① 获取数据源。数据源是 LINQ 查询的对象，可以将数组、集合等作为数据源。例如，将数组 numbers 作为数据源：

```
int[ ] numbers = { 5, 12, 8, 20, 25, 15, 18, 16, 13 };
```

② 创建查询。用于定义查询表达式，通过查询表达式指定如何从数据源中检索信息，并对其排序、分组和结构化。创建查询的一般格式如下：

```
var 查询变量 = from 迭代变量 in 数据源
where...
select...
group 迭代变量或其表达式 by 分组的键
orderby 排序表达式 [ascending][descending]
```

其中，查询变量一般是一个匿名类型的变量，from 子句指定数据源，where 指定查询的条件，select 子句指定返回元素的类型，group by 用于数据分组，orderby 对返回结果排序。查询表达式中 from...select...是必不可少的，其他的可选。

在 LINQ 中，查询变量如果不使用匿名类型的变量，匹配的集合只能使用 IEnumerable<T>接口暴露，其中 T 为返回序列中的数据类型。上面的查询语句也可以写为：

```
IEnumerable<int> subset = from i in numbers where i % 2 == 0 orderby i select i;
```

或者

```
System.Collections.IEnumerable subset = from i in numbers where i % 2 == 0 orderby i select i;
```

③ 执行查询。LINQ 的查询变量本身只是存储查询命令，创建查询仅仅声明了查询变量，并不执行查询的任何操作，不会返回任何数据，只有执行查询后才能返回查询结果，称之为延迟执行。下面的代码执行查询，输出偶数：

```
foreach (int bb in subset)
{
    Response.Write("偶数有: " + "</br>" + bb);
}
```

本章需要使用 studentData 和 scoreData 类，其定义见 6.5.1 小节。

例 10-2 基于 studentData 类，定义集合 stuList 如下，使用 LINQ 查询出"女"生记录，输出 student_no 和 student_name。

```
List<studentData> stuList = new List<studentData>
{
    new studentData{student_no = "20100405",student_name = "张三丰",
        gender = "男",identification = "110140200103083152"},
    new studentData{student_no = "20110607",student_name = "李小梅",
        gender = "女",identification = "110150199005103241"},
    new studentData{student_no = "20120304",student_name = "赵玉霞",
        gender = "女",identification = "110130199206093627"},
```

```
    new studentData{student_no = "20131402",student_name = "张美娟",
      gender = "女",identification = "110140199302075469"}
};
```

使用 LINQ 完成查询和显示的代码：

```
var stuQuery = from ss in stuList where ss.gender == "女"
select new {ss.student_no,ss.student_name };
Response.Write("女生有：<br>");
foreach (var aa in stuQuery)
{
    Response.Write("学号:" + aa.student_no + ",姓名： " + aa.student_name + "<br>");
}
```

说明：本例题查询的数据源的类型是 List<studentData>，查询表达式中输出有两个属性，即 student_no 和 student_name，查询表达式的用法与例 10-1 不同，当查询输出的属性多于一个时，可使用 new，将要输出的内容表示为对象的形式。

10.1.2　LINQ 查询表达式

所有 LINQ 查询表达式都必须有一个指定数据源的 from 子句和表示要获得数据的 select 子句。下面是一个简单的 LINQ 查询，从 stuList 集合中获取女生的记录：

```
var stuQuery = from ss in stuList
               where ss.gender == "女"
               select ss;
```

1. 投影

将正在查询的数据转换为各种结构的过程，称为投影。如果 select 要获得的是数据源中的 student_name，可以修改 select ss 为 select ss. student_name。如果要获取 student_name 和 student_no，一种方法是可以使用 C# 的操作符修改要输出的字符串，如 select student_name+student_no；另一种方法是在 select 子句中添加一个 new 关键字，将要输出的内容以对象的形式赋予属性：

```
var stuQuery = from ss in stuList
                 where ss.gender == "女"
                 select new(学号 = student_no,姓名 = student_name);
```

这个表达式在执行时返回一组使用隐式创建类的对象，该对象有两个属性，即学号和姓名，运行结果见图 10-1。

LINQ 表达式返回一个迭代器对象，迭代器类是泛型。由于使用 select new(sno=student_no,sname=student_name)；时没有定义这个类，故不能正确地定义 IEnumerable<T>的引用(不知道 T 的数据类型)，但查询的结果可以直接使用：

```
女生有：
学号:20110607，姓名：李小梅
学号:20120304，姓名：赵玉霞
学号:20131402，姓名：张美娟
```

图 10-1　LINQ 投影结果

```
var stuQuery = from ss in stuList
             where ss.gender == "女"
             select new { sno = ss.student_no, sname = ss.student_name};
```

也可以将查询的结果 stuQuery 直接绑定到数据控件,如 GridView1,运行结果如图 10-2 所示:

```
GridView1.DataSource = stuQuery;
GridView1.DataBind();
```

sno	sname
20110607	李小梅
20120304	赵玉霞
20131402	张美娟

图 10-2 LINQ 结果直接绑定到 GridView

如果一定要使用 IEnumerable<T>替代 var stuQuery,需要先定义一个 studentInfo 类:

```
public class studentInfo
{
    public string sno
    {set; get;}
    public string sname
    {set; get;}
}
```

查询表达式就可以改为:

```
IEnumerable<studentInfo> stuQuery = from ss in stuList
    where ss.gender == "女"
    select new studentInfo {sno = ss.student_no, sname = ss.student_name};
```

图 10-2 中没有使用 foreach 循环,也可以直接将查询的结果绑定到 GridView 控件上,因为调用 GridView.DataBind()方法时,ASP.NET 对匹配的集合进行迭代后将获得的数据传递给 GridView,此过程与手工迭代的方式触发 LINQ 表达式的计算是相同的。

2. 筛选和排序

LINQ 中使用 where 子句使查询结果过滤为只包括特定条件的记录。下面的代码是从 stuList 中查询姓"张"的记录:

```
var stuQuery = from ss in stuList
        where ss.student_name.StartsWith("张")
        select new {ss.student_no,ss.student_name};
```

在书写筛选条件时,可使用逻辑与(&&)、或(||)操作符及关系运算符(>、<、>=、<=、!=、==)。LINQ 表达式中允许调用自己编写的方法。

例 10-3 查询年龄大于 24 岁的女学生的记录,可写定义一个函数 checkAge(),根据身份证号码判断年龄是否大于 24 岁:

```
private bool checkAge(studentData sDB)
{
    int year = Convert.ToInt16(sDB.identification.Substring(6, 4));
    int month = Convert.ToInt16(sDB.identification.Substring(10, 2));
    int day = Convert.ToInt16(sDB.identification.Substring(12, 2));
    int age = DateTime.Now.Year - year;
    if (DateTime.Now.Month < month || (DateTime.Now.Month == month && DateTime.Now.Day < day))
    {
        age--;
```

```
        }
        if (age > 24)
            return true;
        else
            return false;
}
```

然后定义查询表达式：

```
IEnumerable < studentInfo > stuQuery = from ss in stuList
where ss.gender == "女" && checkAge(ss)
select new studentInfo { sno = ss.student_no, sname = ss.student_name };
```

查询结果如果要排序，可以使用 orderby 排序字段 descending|ascending，默认排序方式是 ascending，下面是一个排序表达式：

```
var stuQuery = from ss in stuList
where ss.gender == "女" && checkAge(ss) orderby ss.student_name descending
select new {sno = ss.student_no, sname = ss.student_name};
```

3. 分组和聚合

将数据源中的信息，根据分组条件建立分组，显示分组后的结果。下面是根据 gender 分组的代码（运行结果见图 10-3）：

图 10-3 分组统计结果

```
var stuQuery = from ss in stuList
    group ss by ss.gender into genderGroup
    select new {gender = genderGroup.Key, count = genderGroup.Count()};
GridView1.DataSource = stuQuery;
GridView1.DataBind();
```

使用 intogenderGroup 命名分组名为 genderGroup，分组得到的结果又是一个集合，该集合根据 Key 键划分每个分组的对象。输出查询结果时，可以使用 LINQ 的聚合运算符，如 Count、Max、Min、Sum、Average 等。

4. 连接

如果需要查询的内容来自多个数据源，两个数据源间一般具有相同的元素，通过此相同元素，将两个数据源连接在一起。与 SQL 类似，这种连接分为左关联、右关联和内部关联。

例 10-4 以 stuList 和 scoreList 两个数据源为例，说明内部关联的使用方法（运行结果见图 10-4）。

```
List < scoreData > scoreList = new List < scoreData >
{
    new scoreData{student_no = "20100405",course_no = "k1",score = 80},
    new scoreData{student_no = "20100405",course_no = "k2",score = 90},
    new scoreData{student_no = "20100405",course_no = "k3",score = 75},
    new scoreData{student_no = "20110607",course_no = "k1",score = 50},
    new scoreData{student_no = "20110607",course_no = "k2",score = 60},
    new scoreData{student_no = "20110607",course_no = "k3",score = 40},
    new scoreData{student_no = "20120304",course_no = "k1",score = 92},
    new scoreData{student_no = "20120304",course_no = "k2",score = 86},
```

```
            new scoreData{student_no = "20131402",course_no = "k1",score = 72},
            new scoreData{student_no = "20131402",course_no = "k2",score = 78},
    };
    var stuQuery = from aa in stuList
                   from bb in scoreList
                   where aa.student_no == bb.student_no
    select new{aa.student_no, aa.student_name, bb.course_no,bb.score};
        GridView1.DataSource = stuQuery;
        GridView1.DataBind();
```

student_no	student_name	class_no	score
20100405	张三丰	k1	80
20100405	张三丰	k2	90
20100405	张三丰	k3	75
20110607	李小梅	k1	50
20110607	李小梅	k2	60
20110607	李小梅	k3	40
20120304	赵玉霞	k1	92
20120304	赵玉霞	k2	86
20131402	张美娟	k1	72
20131402	张美娟	k2	78

图 10-4　两个数据源内部关联运行的结果

10.1.3　LINQ 的立即执行

LINQ 查询表达式只有执行查询时（如用 foreach 循环）才运行，当外部调用 LINQ 查询表达式时，希望查询表达式能够立即执行。通过调用由 Enumerable 类型定义的许多扩展方法来完成。Enumerable 定义了诸如 ToArray<T>()、ToDictionary<TSource, TKey>()及 ToList<T>()在内的许多扩展方法。在调用这些扩展方法的同时将执行 LINQ 查询，以获取数据快照。

```
int[] numbers = { 5, 12, 8, 20, 25, 15, 18, 16, 13 };
//立即获取数据为 int[]
int[] subset = (from i in numbers where i % 2 == 0 orderby i select i).ToArray<int>();
//立即获取数据为 List[int]
List<int> subset = (from i in numbers where i % 2 == 0 orderby i select i).ToList<int>();
```

在调用这些扩展方法时，整个 LINQ 表达式要用圆括号括起来，才能将它强制转换为正确的实际类型来调用 Enumerable 的扩展方法。由于 C♯ 编译器可以准确地检测泛型项的类型参数，故在使用上述扩展方法时，可以不指定类型参数，直接输入：

```
int[] subset = (from i in numbers where i % 2 == 0 orderby i select i).ToArray();
```

或者

```
List<int> subset = (from i in numbers where i % 2 == 0 orderby i select i).ToList();
```

10.1.4　方法查询

LINQ 查询中大多数查询都是使用 LINQ 声明式查询语法，在编译时 C♯ 编译器将所有 C♯ 这些 LINQ 操作符都翻译为对 Enumerable 类中方法的调用，不怕麻烦的话，也可以使用实际的对象模型来构建 LINQ 语句。

Enumerable 的许多方法的原型都是将委托(delegate)作为参数,特别是很多方法都要求一个名为 Func< >的泛型委托作为参数,如 LINQ 查询表达式中的 Where,将调用 Enumerable 的 Where()方法。由于 Enumerable 的许多方法都要求将委托作为输入参数,故调用这些方法时,可以使用委托参数形式,也可以使用 Lambda 表达式,下面是本章开始的示例,如果用方法查询,采用 Lambda 表达式的代码如下:

```
string[] words = { "Vietnam", "Japan", "American", "Korea", "India", "Australia", "France" };
var subset = words.Where(aa => aa.Length > 5).OrderBy(aa => aa);
//var subset = words.Where(aa => aa.Length > 5).OrderBy(bb => bb).Select(cc => cc);
foreach (string bb in subset)
{
    Response.Write(bb + "</br>");
}
```

由于 Lambda 表达式是对于操作匿名方法的简化符号,故上面的代码也可以这样写:

```
string[] words = {"Vietnam", "Japan", "American", "Korea", "India", "Australia", "France"};
Func<string, bool> filter = delegate(string num) {return num.Length > 5;};
Func<string, string> item = delegate(string s) {return s;};
var subset = words.Where(filter).OrderBy(item).Select(item);
foreach (string bb in subset)
{
    Response.Write(bb + "</br>");
}
```

10.2　LINQ to XML

XML(eXtensive Markup Language,可扩展标记语言)是一种标记语言,与 HTML 不同的是它没有被预定义,用户需要自行定义标记。XML 与 HTML 最大的区别是:XML 用于存储和传输数据;HTML 用于显示数据。

例 10-5　studentXML.xml 文件的内容如图 10-5 所示。将该文件内容显示在 GridView 控件中,效果如图 10-6 所示。

程序代码如下:

```
using System.Xml.Linq;
XElement root = XElement.Load(Server.MapPath("studentXML.xml"));
var queryA = from a in root.Elements("student")
             select new
             {学号 = (string)a.Element("student_no"),
              姓名 = (string)a.Element("student_name"),
              性别 = (string)a.Element("gender")
             };
GridView1.DataSource = queryA;
GridView1.DataBind();
```

```xml
<?xml version="1.0" encoding="utf-8"?>
<Students>
  <student>
    <student_no>41321059</student_no>
    <student_name>李孝诚</student_name>
    <gender>男</gender>
    <address>北京市海淀区学院路50号701</address>
    <email>lxc@163.com</email>
    <telephone>13545678934</telephone>
    <identification_ID>110106199802030013</identification_ID>
    <id>1</id>
  </student>
  <student>
    <student_no>41340136</student_no>
    <student_name>马小玉</student_name>
    <gender>女</gender>
    <address>北京市海淀区学院路20号20-102</address>
    <email>mxy@126.com</email>
    <telphone>13545678928</telphone>
    <id>2</id>
  </student>
  <student>
    <student_no>41355062</student_no>
    <student_name>王长林</student_name>
    <gender>男 </gender>
    <address>天津静海30号</address>
    <email>wcl@163.com</email>
    <telephone>13645678934</telephone>
    <identification_ID>110106199512035012</identification_ID>
    <id>3</id>
  </student>
</Students>
```

图 10-5　XML 文件内容

学号	姓名	性别
41321059	李孝诚	男
41340136	马小玉	女
41355062	王长林	男

图 10-6　GridView 中显示 XML

说明：

① LINQ to XML 包含多个类，位于 System.Xml.Linq 命名空间，其中最重要的 3 个类是 XDocument、XElement 和 XAttribute。XDocument 类表示一个 XML 文档，常用的属性和方法见表 10-1。XElement 类表示一个 XML 元素，常用的属性和方法见表 10-2。XAttribute 类表示一个 XML 属性，常用的属性和方法见表 10-3。

表 10-1　XDocument 的常用属性

类别	名　称	说　　明
属性	FirstNode	获取当前节点的第一个子节点
	LastNode	获取当前节点的最后一个子节点
	NextNode	获取当前节点的下一个同级节点
	NodeType	获取当前节点的类型
	Parent	获取此 XObject 的父级 XElement，XObject 表示 XML 树中的节点或者属性
	PreciousNode	获取此节点的上一个同级节点
	Root	获取此文档的 XML 树的根元素
方法	Load	从文件创建一个新的 XDocument
	Save	将一个 XML 文档保存到 XML 文件中

表 10-2 XElement 的常用属性和方法

类别	名称	说明
属性	FirstAttribute	获取当前元素的第一个属性
	FirstNode	获取当前节点的第一个子节点
	LastNode	获取当前节点的最后一个节点
	LastAttribute	获取当前元素的最后一个属性
	Name	获取或者设置元素的名称
	NextNode	获取当前节点的下一个同级节点
	Root	获取此文档的 XML 树的根元素
	Parent	获取此 XObject 的父级 XElement,XObject 表示 XML 树中的节点或者属性
	PreciousNode	获取此节点的上一个同级节点
	Value	获取或者设置此元素的串联文本内容
方法	Load	从文件加载 XElement
	Element	获取具有指定 XName 的第一个(按文档顺序)子元素,如果没有元素具有指定的名称,则返回 NULL,其中 XName 表示 XML 元素或者属性的名称
	Elements	按文档顺序返回此元素或者文档的子元素集合

表 10-3 XAttribute 的常用属性和方法

类别	名称	说明
属性	NextAttribute	获取父元素的下一个属性
	NodeType	获取此节点的类型
	Parent	获取此 XObject 的父级 XElement,XObject 表示 XML 树中的节点或者属性
	PreciousAttribute	获取父元素的上一个属性
	Value	获取或设置此属性的值
方法	Remove	从此属性的父元素中移除
	SetValue	设置此属性的值

② XElement root=XElement.Load(Server.MapPath("studentXML.xml"));从文档 studentXML.xml 加载 XElemment,加载后 root 为该 XML 的根元素,即 Students。由表 10-1 和表 10-2 可以看出,此语句与下面两句效果相同:

```
XDocument doc = XDocument.Load(Server.MapPath("studentXML.xml"));
XElement root = doc.Root;
```

先建立一个表示 XML 的文档对象,然后将其根元素赋值到 root。

例 10-6 将图 10-5 中 id=2 的姓名更改为"张小玉"。

```
XElement root = XElement.Load(Server.MapPath("studentXML.xml"));
IEnumerable<XElement> students = from aa in root.Elements("student")
            where aa.Element("id").Value == "2"
            select aa;
if (students.Count() > 0)
{
```

```
            XElement student = students.First();
            student.SetElementValue("student_name", "张小玉");
            root.Save(Server.MapPath("studentXML.xml"));
        }
```

说明：程序先找到 XML 的根节点，然后使用 LINQ 查询到 id＝2 的所有的 student 元素，将第 1 个 student 元素下的 student_name 更改为"张小玉"，最后保存文件。

例 10-7 增加一条学生记录，各节点及内容如下：

```
< student >
    < student_no > 42401007 </student_no >
    < student_name >赵欣</student_name >
    < gender >女 </gender >
    < address >上海大学 30－202 </address >
    < email > zx@sohu.com </email >
    < telephone > 13345678952 </telephone >
    < id > 6 </id >
</student >
```

代码如下：

```
XElement root = XElement.Load(Server.MapPath("studentXML.xml"));
XElement student = new XElement
(
    "student",
    new XElement("student_no", "42401007"),
    new XElement("student_name", "赵欣"),
    new XElement("gender", "女"),
    new XElement("address", "上海大学 30－202"),
    new XElement("email", "zx@sohu.com"),
    new XElement("telephone", "13345678952"),
    new XElement("id", "6")
);
root.Add(student);
root.Save(Server.MapPath("studentXML.xml"));
```

10.3 LINQ to Entities

　　LINQ to Entities 允许开发人员使用 C♯等给实体框架概念模型编写查询。LINQ to Entities 最终会创建在后台数据库中执行的 SQL，而 LINQ to Entities 将 LINQ 查询转换为 ADO.NET 实体框架（Entity FrameWork,EF）能理解的命令查询，对实体框架执行这些查询，返回同时由实体框架和 LINQ 使用的对象。

　　微软把开发的重点已经从 LINQ to SQL 转移到 LINQ to Entities，不支持 LINQ to SQL 的更新。Entity Framework 和 LINQ to Entities 与 LINQ to SQL 的使用相似,但额外的数据库支持以及一些高级模型功能是使用 LINQ to SQL 不可能实现的。微软建议编程时使用 LINQ to Entities 取代 LINQ to SQL。

10.3.1 生成数据模型

Entity Framework 依赖于一个数据模型来使用 LINQ to Entities 进行查询。表中的行被转换为 C♯ 对象的实例,表中的列是这些对象的属性。数据库架构和数据模型对象的映射是 Entity Framework 的核心,是 LINQ to Entities 工作的根本。

多数开发人员都会使用 Visual Studio 自动建立数据模型。建立实体对象模型的方法如下。

❶ 右击"解决方案资源管理器",选择快捷菜单中的"添加新项"命令,在项目模板中选择"ADO.NET 实体数据模型",设置要创建的文件名称(如 StudentModel.edmx),如图 10-7 所示,单击"添加"按钮。

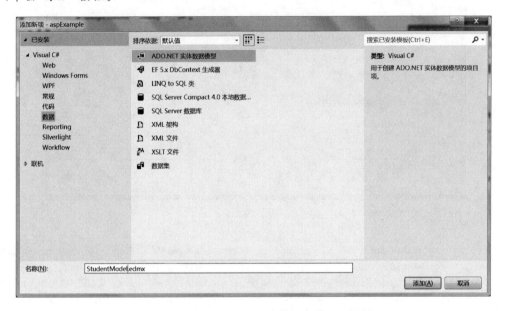

图 10-7 "添加 ADO.NET 实体对象模型"对话框

❷ 虽然可以选择"空模型",然后手工增加类,但目前要从一个已经存在的数据库 students 中生成模型。在"实体数据模型向导"对话框中选择"从数据库生成"并配置数据库连接,如图 10-8 所示。可以选择要包含在数据模型中的数据库表、视图和存储过程。Visual Studio 随即会为选择的数据库创建模型图,它显示了已创建的映射对象、对象拥有的字段以及对象间的关系。图 10-8 中将实体连接的设置另存为 studentEntities,单击"完成"按钮后项目中会增加以下文件。

StudentModel.edmx:是一个 XML 文件,用于定义数据库模型的构架。

StudentModel.Designer.cs:是一个 C♯代码文件,包含用于 LINQ to Entities 查询的数据类型。

通常 StudentModel.Designer.cs 的内容是隐藏的,而且这个文件会被数据模型重新生成,故不应该也没有必要去修改这个文件。

上述操作将 students 数据库的数据模型映射为对象模型,所创建的对象模型为上下文类 studentEntities,并在 Web.config 文件的<connectionStrings>节中自动添加连接字符串:

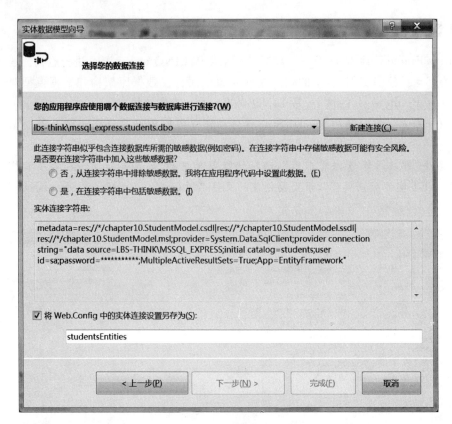

图 10-8 建立数据库连接

< add name = "studentsEntities" connectionString = "metadata = res:// * /chapter10. StudentModel. csdl | res:// * /chapter10. StudentModel. ssdl | res:// * /chapter10. StudentModel. msl; provider = System. Data. SqlClient; provider connection string = " data source = LBS - THINK\ MSSQL_EXPRESS; initial catalog = students; user id = sa; password = asp123; MultipleActiveResultSets = True; App = EntityFramework"" providerName = "System. Data. EntityClient" />

上下文区域定义的第一个类从 ObjectContext 派生,其名称为 studentEntities,这个类的构造函数连接到生成模型的数据库,或者可以指定连接字符串连接到其他数据库(需要具有相同的架构;否则模型无法工作)。

10.3.2 LINQ to Entities 查询

例 10-8 使用 LINQ to Entities 从 student 中筛选出"女生"并按照姓名排序。

新建 10_5. aspx 文件,在页面中增加 GridView1 控件和命令按钮 Button1 各一个,在 Button1 中写入以下代码:

```
protected void Button1_Click(object sender, EventArgs e)
{
    studentsEntities dc = new studentsEntities();
    var varQuery = from m in dc.student
                   where m.gender == "女"
                   orderby m.student_name
```

```
            select m;
    GridView1.DataSource = varQuery.ToList();
    //GridView1.DataSource = varQuery.ToArray();
    GridView1.DataBind();
}
```

代码首先创建一个 studentsEntities 的实例 dc，隐式地创建了与数据库的连接。其次在查询时 LINQ 使用了 student 属性作为数据源。需要注意的是，varQuery 是一个存储查询，其值不是一个集合，需要由 Enumerable 类型定义的许多扩展方法如 ToArray<T>()、ToDictionary<TSource,TKey>()及 ToList<T>()运行查询后，才能生成集合。

例 10-9 按性别统计 student 中人数。

在例 10-8 的 10_5.aspx 文件中，增加一个命令按钮 Button2，在 Button2 中写入以下代码：

```
protected void Button2_Click(object sender, EventArgs e)
{
    studentsEntities dc = new studentsEntities();
    var varQuery = from m in dc.student
                group m by m.gender into genderGroup
                select new
                {
                    性别 = genderGroup.Key,
                    人数 = genderGroup.Count()
                };
    GridView1.DataSource = varQuery.ToList();
    GridView1.DataBind();
}
```

本例使用了聚合函数 Count()。

例 10-10 按学生的选课门数统计，运行结果如图 10-9 所示。

在例 10-8 的 10_5.aspx 文件中，增加一个命令按钮 Button3，在 Button3 中写入以下代码：

```
protected void Button3_Click(object sender, EventArgs e)
{
    studentsEntities dc = new studentsEntities();
    var varQuery = from st in dc.student
        let sc = from sc in dc.score where st.student_no == sc.student_no
        select sc
        select new{学号 = st.student_no,姓名 = st.student_name,选课门数 = sc1.Count()};
    GridView1.DataSource = varQuery.ToList();
    GridView1.DataBind();
}
```

学号	姓名	选课门数
30012035	楚云飞	1
40012030	赵三	1
41321059	李孝诚	3
41340136	马小玉	4
41355062	王长林	3
41361045	李将寿	3
41361258	刘登山	0
41361260	张文丽	0
41401007	鲁宇星	3
41405002	王小月	3
41405003	张晨露	3
41405005	张伟达	0
41405221	张康林	0
41405238	张雨	1

图 10-9 统计学生选课门数

实体包含导航属性，通过导航属性可以在数据模型间移动而不需要考虑外键关系。上面的查询就使用了导航属性。st 是 student 对应的数据集，由于 student 表与 score 表之间是一对多的关系，故通过 st 可以导航到 score。

10.3.3　LINQ to Entities 数据库操作

1. 插入记录

插入记录时首先要建立一个要插入的记录，再用 Add 添加到集合中，然后调用 SaveChanges 方法提交插入。代码如下：

```
studentsEntities dc = new studentsEntities();
student st = new student {
    student_no = "B20150102",
    student_name = "郑伟",
    email = "abc@163.com",
    identification = "1204051990102O3461",
    gender = "女"
};
dc.student.Add(st);
dc.SaveChanges();
```

2. 更新记录

需要先找到要更新的记录，然后修改相应的字段值，最后调用 SaveChanges 方法完成更新。代码如下：

```
studentsEntities dc = new studentsEntities();
student st = (from c in dc.student where
              c.student_no == "B2015010" select c).Single();   //查找要更新的记录
st.student_name = "郑伟伟";
st.gender = "女";
dc.SaveChanges();
```

3. 删除记录

需要先找到要删除的记录，然后用 Remove 方法从集合中删除该记录，最后再调用 SaveChanges 方法完成更新。代码如下：

```
studentsEntities dc = new studentsEntities();
//使用 lambda 查找到学号 B2015010;
student st = dc.student.Single(m => m.student_no == "B2015010");
dc.student.Remove(st);
dc.SaveChanges();
```

习题与思考

（1）从 Light Red、Green、Yellow、Dark Red、Red、Purple 中找到带有 Red 的单词。

（2）finishers.xml 文件中的内容如下，请用 LINQ to XML 技术读取数据并显示在 GridView 中：

```
<?xml version = "1.0" encoding = "utf-8"?>
<finishers>
```

```
<runner>
    <fname>John</fname>
    <lname>Smith</lname>
    <gender>m</gender>
    <time>25:31</time>
    <pic></pic>
</runner>
<runner>
    <fname>Mary</fname>
    <lname>Brown</lname>
    <gender>f</gender>
    <time>26:01</time>
    <pic></pic>
</runner>
...
</finishers>
```

(3) 通过 LINQ 查询表达式从数字集合{10,5,8,16,18,25,40}中获取大于 15 的数显示在网页上。

(4) 使用 LINQ to Entities 查询出 teacher 表中所有记录,并显示在 GridView 上。

第 11 章

ASP.NET Web 服务

(1) 建立 ASP.NET Web 服务。
(2) 使用 ASP.NET Web 服务。
(3) 使用 Newtonsoft.Json,创建提供 JSON 格式的 ASP.NET Web 服务。

11.1 Web 服务的应用

Web 服务是一种无须购买并部署的组件,这种组件是被一次部署到 Internet 中,然后到处可以使用的一种新型组件,只要能够接入 Internet,就可以使用和集成 Web 服务。Web 服务的应用目前非常广泛,可以使用网络上已经存在的 Web 服务,也可以从数据库服务器中提取数据供 Web 服务者使用。在企业内部的局域网中,内部员工可以任意调用公司内部的服务资源,简化自己的工作。本节实例介绍查询火车列车时刻表、天气预报等 Web 服务的应用。本章示例存放在 aspExample\chapter11 文件夹下。

例 11-1 借助提供火车运行时刻表的 Web 服务,通过选择发车站和到车站,查询相关火车运行信息,以列表的形式显示在 GridView 控件中。

❶ 在"解决方案资源管理器"中右击"引用",在弹出的快捷菜单中选择"添加服务引用"命令,打开"添加服务引用"对话框,单击"高级"按钮,出现"服务引用设置"对话框,单击"添加 Web 引用"按钮,弹出"添加 Web 引用"对话框,在该对话框中输入:

http://www.webxml.com.cn/WebServices/TrainTimeWebService.asmx

单击输入框后面表示"转到"的按钮 ➡,图 11-1 中显示出相关调用方法及其说明。

❷ 单击 ➡ 按钮后,在图 11-1 下面的服务详情窗口中共列出了 8 个获取火车运行时刻的操作。

下面介绍常用的两个。

(1) GetStationAndTimeByStationName:根据发车站和到达站查询火车时刻表。输入

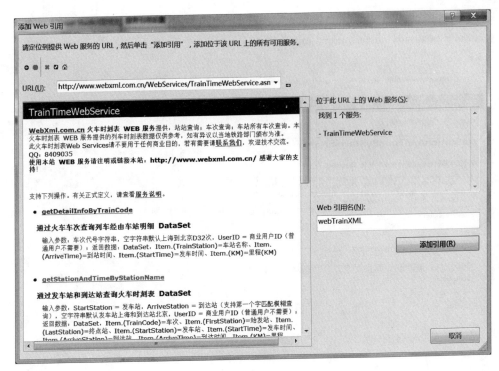

图 11-1 "添加 Web 引用"对话框

的参数:StartStation=发车站,ArriveStation=到达站(支持第一个字匹配模糊查询),空字符串默认发车站上海,到达站北京,UserID=商业用户 ID(普通用户不需要);返回数据:DataSet,Item.(TrainCode)=车次、Item.(FirstStation)=始发站、Item.(LastStation)=终点站、Item.(StartStation)=发车站、Item.(StartTime)=发车时间、Item.(ArriveStation)=到达站、Item.(ArriveTime)=到达时间、Item.(KM)=里程 KM、Item.(UserDate)=历时。

(2) getSationName:获得本火车时刻表 Web Services 提供的全部始发站的名称,输入参数:无;输出参数:字符串类型的数组。

在应用程序中根据这两个操作的要求编写对应调用的方法即可。可以测试该服务是否能够正常使用,单击 getStationAndTimeByStationName,显示图 11-2 所示的对话框,输入发车站和到达站,单击"调用"按钮,浏览器中以 XML 格式显示出列车相关信息。

❸ 在图 11-1 或图 11-2 中输入 Web 引用名 webTrainXML,单击"添加引用"按钮。系统会在当前项目中建立一个文件夹 Web References 及其子文件夹 webTrainXML。

❹ 新建 Web 窗体 11_1.aspx,在页面上拖放两个 DropDownList 控件,设置其 ID 分别为 DropFrom 和 DropTo;拖放一个命令按钮并设置其 ID 为 BtnSearch;拖放一个 GridView 控件 GridView1。

❺ 在 Page_Load 中写入如下代码:

```
protected void Page_Load(object sender, EventArgs e)
{
    webTrainXML.TrainTimeWebService myService = new webTrainXML.TrainTimeWebService();
    string[] strStation = myService.getStationName();
    for (int i = 0; i < strStation.Length; i++)
```

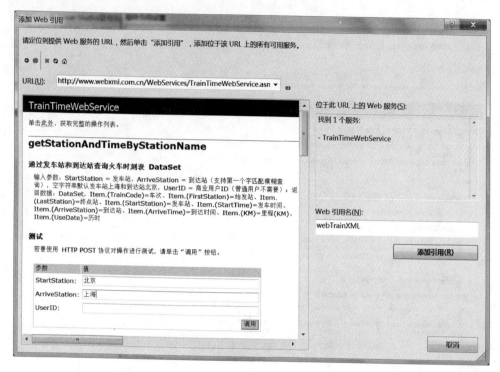

图 11-2 "添加 Web 引用"对话框中测试 Web Services

```
{
    DropFrom.Items.Add(strStation[i]);
    DropTo.Items.Add(strStation[i]);
}
}
```

说明：代码中的 webTrainXML 就是图 11-1 ❷ 中命名的 Web 引用名。调用 getStationName()方法得到本服务提供的所有发车站站名，将其加入到用 DropDownList 表示的发车站和到车站列表中。

❻ "查询"按钮 BtnSearch 下的代码：

```
protected void BtnSearch_Click(object sender, EventArgs e)
{
    webTrainXML.TrainTimeWebService myService = new webTrainXML.TrainTimeWebService();
    GridView1.DataSource = myService.getStationAndTimeByStationName(DropFrom.Text,DropTo.Text,"");
    GridView1.DataBind();
}
```

❼ 运行 11_1.aspx 文件，结果如图 11-3 所示。

例 11-2 借助提供天气预报的 Web Service，通过选择省份和相应的城市来获取该城市的天气预报信息。

起始站：	北京 ▼	终点站	大同 ▼	查询				
TrainCode	FirstStation	LastStation	StartStation	StartTime	ArriveStation	ArriveTime	KM	UseDate
1133\1136	天津	乌海西	北京西	23:02:00	大同	05:21:00	368	06:19
2602\2603	秦皇岛	大同	北京	00:18:00	大同	06:31:00	374	17:47
K217\K220	邯郸	银川	北京西	00:33:00	大同	06:47:00	368	06:14
K41\K44	北京	敦煌	北京	21:40:00	大同	04:40:00	374	07:00
K573	北京西	东胜西	北京西	20:35:00	大同	02:48:00	368	06:13
K597\K600	广州	包头	北京西	21:26:00	大同	03:30:00	368	17:56
K615	北京	大同	北京	15:33:00	大同	22:11:00	374	06:38
K695	北京	大同	北京	23:32:00	大同	06:05:00	374	06:33
K960\K961	沈阳	临汾	北京	02:49:00	大同	09:06:00	374	06:17

图 11-3　列车时刻表查询结果

❶ 在"解决方案资源管理器"中右击"引用",在弹出的快捷菜单中选择"添加服务引用"命令,打开"添加服务引用"对话框,单击"高级"按钮,出现"服务引用设置"对话框,单击"添加 Web 引用"按钮,弹出"添加 Web 引用"对话框,在该对话框中输入:

http://www.webxml.com.cn/WebServices/WeatherWebService.asmx

单击输入框后面表示"转到"的按钮 ➡,图 11-4 中显示出相关调用方法及其说明。

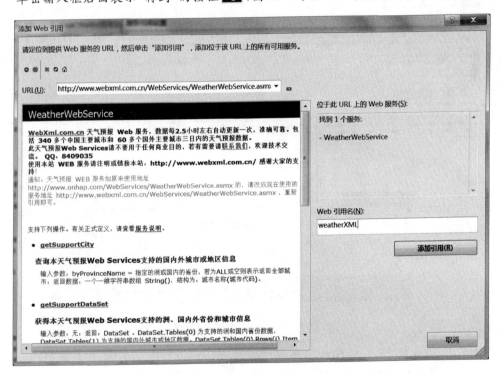

图 11-4　"添加 Web 引用"对话框

❷ 进入服务操作列表窗口,在服务详情中列出了 5 个获取天气预报的相关操作,下面列出本示例中用到的 3 个。

getSupportProvince：获得本天气预报 Web 服务支持的洲、国内外省份和城市信息。输入参数：无；返回数据：一个一维字符串数组 String(),内容为洲或国内省份的名称。

getSupportCity：查询本天气预报 Web Services 支持的国内外城市或地区信息。输入参数：byProvinceName=指定的洲或国内的省份,若为 ALL 或空则表示返回全部城市；返回数据：一个一维字符串数组 String()；结构：城市名称(城市代码)。

getWeatherbyCityName：根据城市或地区名称查询获得未来 3 天内天气情况、现在的天气实况、天气和生活指数。

调用方法如下：输入参数 theCityName＝城市中文名称(国外城市可用英文)或城市代码(不输入默认为上海市)，如上海或 58367，如有城市名称重复请使用城市代码查询(可通过 getSupportCity 或 getSupportDataSet 获得)；返回数据：一个一维数组 String(22)，共有 23 个元素。

应用程序中根据这 3 个操作的要求对应编写调用的方法即可。可以测试该服务器是否能够正常使用：单击 getWeatherbyCityName，进入图 11-5，根据说明在参数 theCityName 城市名称后面的文本框中输入"北京"，单击"调用"按钮，显示图 11-6。

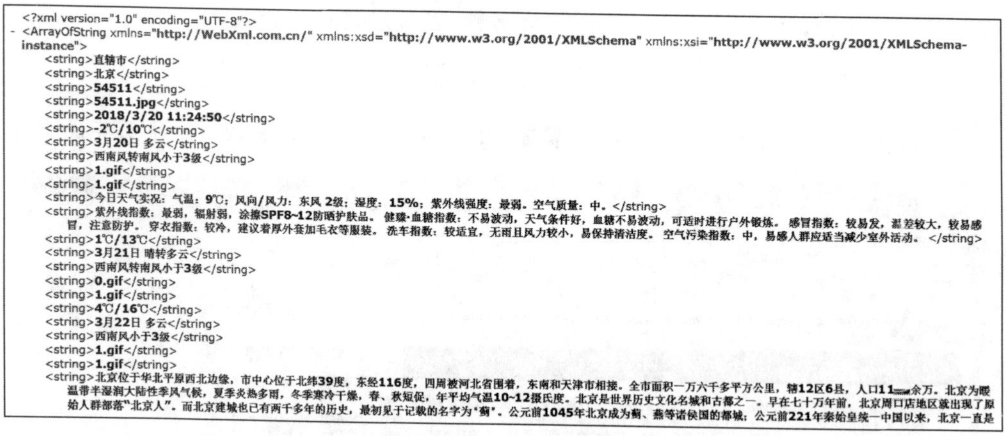

图 11-5　测试天气预报的 Web 服务

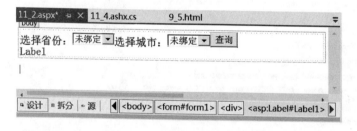

图 11-6　示例 Web 窗体页面布局

❸ 在图 11-4 的"Web 引用名"文本框中输入 weatherXML，单击"添加引用"按钮。

❹ 新建页面 11_2.aspx，在"设计视图"对话框下，从工具箱中拖曳 2 个 DropDownList 控件、1 个 Button 控件，分别修改其 ID 属性为 DropProvince、DropCity、BtnSearch；设置 DropProvince 的 AutoPostBack＝"True"，拖放 1 个 Label 控件 Label1。按照图 11-6 布置页面。

❺ 在 Page_Load 中写入如下代码：

```
protected void Page_Load(object sender, EventArgs e)
{
    if (!Page.IsPostBack)
```

```
        {
            BindToProvince();
            BindToCity();
        }
}
```

❻ 编写 BindToProvince()和 BindToCity(),代码如下:

```
private void BindToProvince()
{
    weatherXML.WeatherWebService myService = new weatherXML.WeatherWebService();
    string[] strProvince = myService.getSupportProvince();
    foreach (string aa in strProvince)
    {
        DropProvince.Items.Add(aa);
    }
}
private void BindToCity()
{
    weatherXML.WeatherWebService myService = new weatherXML.WeatherWebService();
    string[] strCity = myService.getSupportCity(DropProvince.Text);
    foreach (string aa in strCity)
    {
        DropCity.Items.Add(aa.Substring(0,aa.Length - 7));
    }
}
```

❼ 在 DropProvince_SelectedIndexChanged 中编写以下代码:

```
protected void DropProvince_SelectedIndexChanged(object sender, EventArgs e)
{
    DropCity.Items.Clear();
    BindToCity();
}
protected void BtnSearch_Click(object sender, EventArgs e)
{
    Label1.Text = "";
    weatherXML.WeatherWebService myService = new weatherXML.WeatherWebService();
    string[] strWeather = myService.getWeatherbyCityName(DropCity.Text);
    foreach (string aa in strWeather)
    {
        if (aa.Contains("jpg") == false && aa.Contains("gif") == false)
            Label1.Text += aa + "<br/>";
    }
}
```

❽ 运行 11_2.aspx 文件,选择省份后,城市中自动改变为该省份下的城市,单击"查询"按钮后,运行执行结果如图 11-7 所示。

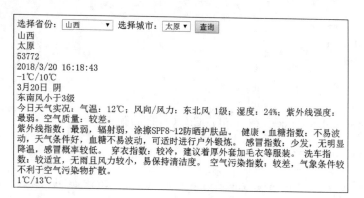

图 11-7 天气预报 Web 服务运行结果

11.2 创建提供查询学生成绩的 Web 服务

除了使用已经存在的 Web 服务外,还可以自己创建提供给其他用户使用的 Web 服务。下面以建立查询学生成绩的 Web 服务为例,说明建立 Web 服务和使用自己建立 Web 服务的过程。

例 11-3 创建提供查询学生成绩的 Web 服务,显示"学号""姓名""课程名""成绩"。

❶ 在"解决方案资源管理器"对话框中右击文件夹 chapter11,在弹出的快捷菜单中选择"添加"→"新建项"命令,选择对话框中的"Web 服务"选项,在"名称"文本框中输入 StudentsWebService.asmx,单击"添加"按钮,如图 11-8 所示。

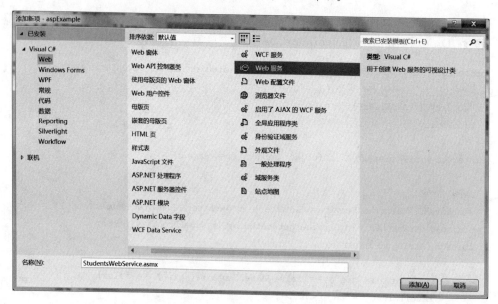

图 11-8 "添加新项"对话框

❷ StudentsWebService.asmx.cs 自动被打开,然后添加以下代码:

```
1  using System;
2  using System.Collections.Generic;
```

```csharp
3   using System.Linq;
4   using System.Web;
5   using System.Web.Services;
6   using System.Data;
7   using System.Data.SqlClient;
8   using System.Web.Configuration;
9   namespace aspExample.chapter11
10  {
        /// <summary>
        /// StudentsWebService 的摘要说明
        /// </summary>
11      [WebService(Namespace = "http://tempuri.org/")]
12      [WebServiceBinding(ConformsTo = WsiProfiles.BasicProfile1_1)]
13      [System.ComponentModel.ToolboxItem(false)]
        // 若要允许使用 ASP.NET Ajax 从脚本中调用此 Web 服务,请取消注释以下行
14      //[System.Web.Script.Services.ScriptService]
15      public class StudentsWebService:System.Web.Services.WebService
16      {
17          [WebMethod]
18          public DataTable GetXMLScore()
19          {
                //提供 XML 格式的学生成绩
20              string cnnString = WebConfigurationManager.ConnectionStrings["SQLServerString"].ConnectionString;
21              SqlConnection conn = new SqlConnection(cnnString);
22              string strSQL = "select student.student_no as 学号,student.student_name as 姓名,course.course_name as 课程名,score.score as 成绩 from student,course,score where student.student_no = score.student_no and score.course_no = course.course_no";
23              SqlDataAdapter adapter = new SqlDataAdapter(strSQL, conn);
                //填充数据集 ds,加入的表名命名为 student
24              DataSet ds = new DataSet();
25              adapter.Fill(ds, "student");
26              DataTable dt = ds.Tables[0];
27              return dt;
28          }
29      }
30  }
```

说明:

① 第 17 行添加了一个名为 WebMethod 的属性,该属性用来标志 GetXMLScore() 方法可以被远程的客户端访问。

② 第 20 行从 Web.config 中读取连接数据库的字符串,需要引入第 8 行的命名空间。

③ 需要查询的"学号""姓名""课程名""成绩"来自 student、course、score 3 个表,第 22 行通过内部关联的方法,组合查询字符串。

❸ 在 VS 中运行 StudentsWebService.asmx,出现图 11-9 所示的运行界面,单击 GetXMLScore 超链接,出现图 11-10 所示的运行界面,单击"调用"按钮,出现图 11-11 所示的运行界面,表示 Web 服务的程序运行正常。示例中调用 GetXMLScore 服务,查询的结果采用 XML 的形式返回。

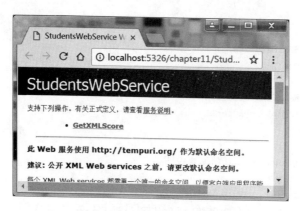

图 11-9　运行 Web 服务的界面

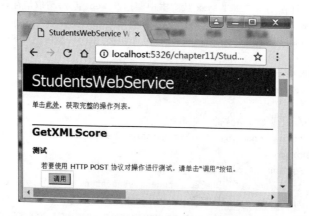

图 11-10　运行 GetXMLScore 的界面

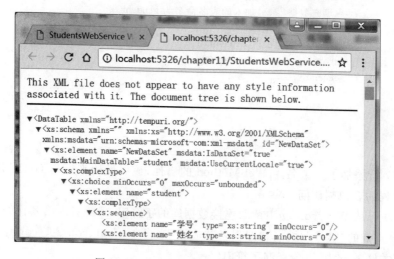

图 11-11　GetXMLScore 服务运行的结果

❹ 在"解决方案资源管理器"对话框中右击"引用",在弹出的快捷菜单中选择"添加服务引用"命令,打开"添加服务引用"对话框,保持默认的命名空间 ServiceReference1,单击"高级"按钮,显示图 11-12。单击"添加 Web 引用"按钮,在出现的对话框的地址栏中填写要

访问的 Web 应用的地址。本示例访问的 Web 应用在本地,如图 11-13 所示,输入的网址是:
http://localhost:81/example/chapter11/StudentsWebService.asmx

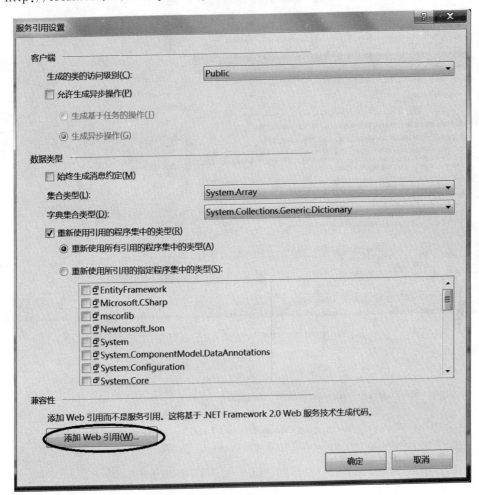

图 11-12 "服务引用设置"对话框

❺ 图 11-13 中"Web 引用名"文本框中输入 localhost,单击"添加引用"按钮,打开 Web.config 文件,看到其中自动添加了以下代码:

```
<applicationSettings>
 <aspExample.Properties.Settings>
    <setting name = "aspExample_localhost_StudentsWebService" serializeAs = "String">
      <value>http://localhost:81/example/chapter11/StudentsWebService.asmx</value>
    </setting>
 </aspExample.Properties.Settings>
</applicationSettings>
```

说明:这表示设定了代理类 localhost 所引用的 Web 服务的 URL。在"解决方案资源管理器"对话框中看到新建了一个 Web References 文件夹,其中包含 localhost 类的子文件夹。

❻ 新建 Web 窗体 11_3.aspx,在窗体上拖放 1 个 GridView 控件 GridView1,写入以下

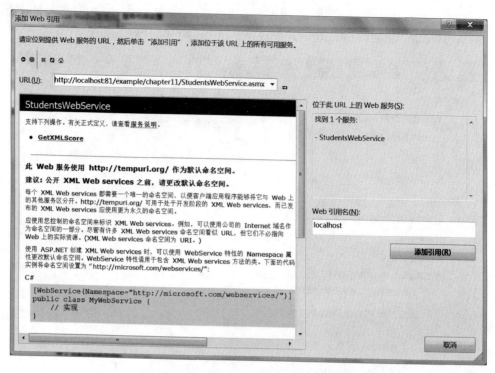

图 11-13 "添加 Web 引用"对话框

代码：

```
protected void Page_Load(object sender, EventArgs e)
{
    localhost.StudentsWebService myServices = new localhost.StudentsWebService();
    GridView1.DataSource = myServices.GetXMLScore();
    GridView1.DataBind();
}
```

❼ 运行 11_3.aspx 文件，显示结果如图 11-14 所示。

图 11-14 使用 Web Service 显示学生成绩

例 11-4　提供查询学生成绩的 Web 服务,要求输出的数据格式是 JSON。

Web 程序开发中,在使用某些框架如 jQuery、Extjs 等时,Ajax 异步请求的数据格式是 JSON,故很有必要提供 JSON 格式的 Web 服务。

❶ 打开例 11-3 的 StudentsWebService.asmx.cs 程序,在程序中增加以下命名空间:

using Newtonsoft.Json;

❷ 在 StudentsWebService 类中增加以下代码:

```
1  [WebMethod]
2  public string GetJsonScore()
3  {
      //提供 JSON 格式的学生成绩
4     return DataTableToJsonWithJsonNet(GetXMLScore());
5  }
6  private string DataTableToJsonWithJsonNet(DataTable table)
7  {
8     string JsonString = string.Empty;
9     JsonString = JsonConvert.SerializeObject(table);
10    return JsonString;
11 }
```

说明:

① 第 1 行添加了一个名为 WebMethod 的属性,该属性用来标志 GetJsonScore()方法可以被远程的客户端访问。

② 第 6～11 行使用 Newtonsoft.Json 将 table 转变为 JSON 格式,除这种方法外,还可以使用其他方法,将 table 转变为指定格式的 JSON,具体可以参考第 9 章。

❸ 在 VS 中运行 StudentsWebService.asmx,在网页中单击超链接 GetJsonScore 后,再单击"调用"按钮,显示图 11-15,输出的数据是数组格式的 JSON 字符串:

[{"学号":"30012035","姓名":"楚云飞 ","课程名":"ASP 网络数据库 ","成绩":88.0},{"学号":"40012030","姓名":"赵三 ","课程名":"ASP 网络数据库 ","成绩":88.0},…]

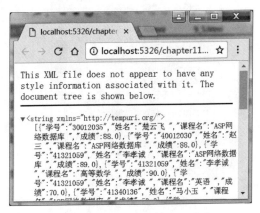

图 11-15　Web 应用输出 JSON

❹ 在 chapter11 文件夹下新建普通网页文件 11_4.html,页面中的代码如下:

```
1  <body>
```

```
 2    <script type="text/javascript" src="../Scripts/jquery.min.js"></script>
 3    <div id="content">
 4    </div>
 5    <script>
 6        $.ajax({
 7            type: "POST",
 8            url: "StudentsWebService.asmx/GetJsonScore",
 9            dataType: "json",
10            contentType: "application/json; charset=utf-8",
11            success: function (json) {
12                var jsonObj = eval('(' + json.d + ')');
13                for (var i = 0; i < jsonObj.length; i++)
14                {
15                    $("#content").append(jsonObj[i].学号 + "," + jsonObj[i].姓名 + "," + jsonObj[i].课程名 + "," + jsonObj[i].成绩 + "<br/>");
16                }
17            },
18            error: function (error) {
19                alert("调用出错" + error.responseText);
20            }
21        });
22    </script>
23    </body>
```

说明：第12行用Ajax获取Web Service中的数据时，由于微软Web应用默认情况下返回JSON数据的格式为{"d":"后台返回的数据"}，返回的数据放在d属性中，故需要使用json.d的方式读取返回的数据；json.d得到返回的JSON字符串，通过eval()将JSON字符串转变为JSON对象。

❺ 运行11_4.html，结果如图11-16所示。

```
30012035,楚云飞 ,ASP网络数据库 ,88
40012030,赵三 ,ASP网络数据库 ,88
41321059,李孝诚 ,ASP网络数据库 ,89
41321059,李孝诚 ,高等数学 ,90
41321059,李孝诚 ,英语 ,70
41340136,马小玉 ,ASP网络数据库 ,50
41340136,马小玉 ,高等数学 ,80
41340136,马小玉 ,英语 ,77
41340136,马小玉 ,人工智能 ,60
41355062,王长林 ,ASP网络数据库 ,90
41355062,王长林 ,高等数学 ,88
41355062,王长林 ,英语 ,75
41361045,李将寿 ,ASP网络数据库 ,66
41361045,李将寿 ,高等数学 ,85
41361045,李将寿 ,英语 ,90
41401007,鲁宇星 ,ASP网络数据库 ,82
41401007,鲁宇星 ,高等数学 ,62
41401007,鲁宇星 ,英语 ,74
41405002,王小月 ,ASP网络数据库 ,90
41405002,王小月 ,高等数学 ,84
41405002,王小月 ,英语 ,85
41405003,张晨露 ,ASP网络数据库 ,81
41405003,张晨露 ,高等数学 ,72
41405003,张晨露 ,英语 ,76
41405238,张雨 ,ASP网络数据库 ,88
```

图11-16 使用Ajax前台显示Web应用后台返回的数据

习题与思考

(1) 利用 Web Service 获取手机号码所在地的信息。该 Web Service 的地址为 http://www.webxml.com.cn/WebServices/MobileCodeWS.asmx。

(2) 借助提供飞机航班时刻表的 Web Service,通过选择起始站、终点站和航班日期,查询相关的航班信息,并显示在列表中,该服务的地址为 http://www.webxml.com.cn/WebServices/domesticairline.asmx。

(3) 验证 Email 地址是否正确,提供服务的网址为 http://www.webxml.com.cn/WebServices/ValidateEmailWebService.asmx,输入一个 email,验证该 email 是否正确。

(4) 建立一个 Web Service,可以提供全部课程的信息。

第 12 章

jQuery EasyUI

（1）jQuery EasyUI 的使用方法。
（2）jQuery EasyUI 页面布局。
（3）jQuery EasyUI 中的消息框。
（4）datagrid 的用法。

12.1 jQuery EasyUI 概述

jQuery EasyUI 是在 jQuery 的基础上制作用户界面的一组控件，借助它可以用少量的 JavaScript 代码开发界面友好、交互性强的 Web 页面。该架构支持 HTML5，简单易学。jQuery EasyUI 的官方网站是 http://www.jeasyui.com/index.php，网站上提供了大量的教程和示例，示例可以在线运行。示例的服务器端代码虽然是 PHP，但其在 ASP.NET 中的用法也可以借鉴。

使用 jQuery EasyUI 前，需要先下载 jQuery EasyUI 的类库。类库中 jQuery 和 jQuery EasyUI 有两种类型的版本：一种是文件名上有 min，供那些只是应用，不准备修改其源代码的用户使用，称之为"生产版本"；另一种是文件名上没有 min，可修改其源代码使用，是开发版本。使用 jQuery 时一定要注意 jQuery 和 jQuery EasyUI 的版本号。本书中使用的 jquery.min.js 版本是 v1.11.3，jquery.easyui.min.js 的版本是 1.5.4，是从官方网站上下载的。用记事本打开这两个文件，里面有版本的说明。下载的压缩文件中包含的内容如图 12-1 所示。

在本地站点建立一个文件夹 Scripts，将下载后的文件全部放在该文件夹下。Scripts 文件夹在站点的位置和要编辑的网页的相对位置，决定了引用 jQuery 类库的路径。本书中各章节的示例放在不同的文件夹下，如第 5 章示例放在 chaper5 中，Scripts 与各章节的相对关系如图 12-2 所示。

图 12-1　下载包中包含的文件

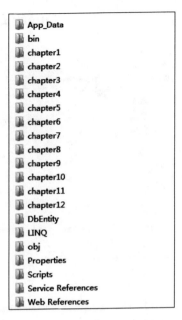

图 12-2　站点 aspExample 下 Scripts 的位置

在各章节的网页中,使用 jQuery EasyUI 引用的代码如下：

```
<head>
<meta http-equiv="Content-Type" content="text/html; charset=utf-8"/>
<title></title>
<link rel="stylesheet" type="text/css" href="../Scripts/themes/default/easyui.css"/>
<link rel="stylesheet" type="text/css" href="../Scripts/themes/icon.css"/>
<link rel="stylesheet" type="text/css" href="../Scripts/demo/demo.css"/>
</head>
```

下面引用的 EasyUI 库文件可以放在头文件中,也可以放在 body 中,建议放在 body 中,以加快网页加载的速度。"../"表示返回到当前文件夹的上一级文件夹。

```
<body>
<script type="text/javascript" src="../Scripts/jquery.min.js"></script>
<script type="text/javascript" src="../Scripts/jquery.easyui.min.js"></script>
</body>
```

12.2　jQuery EasyUI 的 Layout

例 12-1　用 jQuery EasyUI,编写出的网页效果如图 12-3 所示。

新建网页 12_1.html,视图"源"中的内容如下：

```
<!DOCTYPE html>
<html>
<head>
    <meta charset="UTF-8">
```

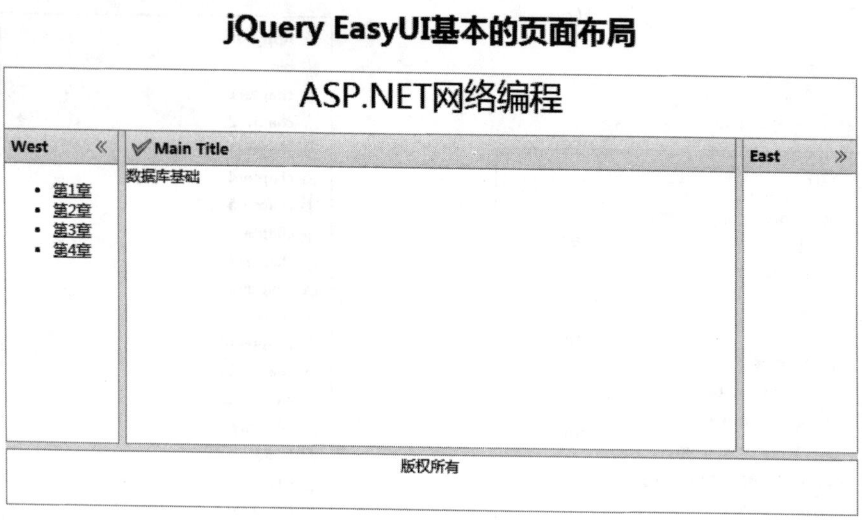

图 12-3 基本页面布局

```html
<title>Basic Layout - jQuery EasyUI Demo</title>
<link rel="stylesheet" type="text/css" href="../Scripts/themes/default/easyui.css">
<link rel="stylesheet" type="text/css" href="../Scripts/themes/icon.css">
<link rel="stylesheet" type="text/css" href="../Scripts/demo.css">
</head>
<body>
    <script type="text/javascript" src="../Scripts/jquery.min.js"></script>
    <script type="text/javascript" src="../Scripts/jquery.easyui.min.js"></script>
    <h2 style="text-align:center">jQuery EasyUI 基本的页面布局</h2>
    <div style="margin:20px 0;"></div>
    <center>
    <div class="easyui-layout" style="width:700px;height:350px;">
    <div data-options="region:'north'" style="height:50px;font-size:20pt;text-align:center">ASP.NET 网络编程</div>
    <div data-options="region:'south',split:true" style="height:50px;text-align:center">版权所有</div>
        <div data-options="region:'east',split:true" title="East" style="width:100px;"></div>
        <div data-options="region:'west',split:true" title="West" style="width:100px;">
          <ul>
            <li><a href="javascript:void(0)" onclick="showcontent('1')">第 1 章</a></li>
            <li><a href="javascript:void(0)" onclick="showcontent('2')">第 2 章</a></li>
            <li><a href="javascript:void(0)" onclick="showcontent('3')">第 3 章</a></li>
            <li><a href="javascript:void(0)" onclick="showcontent('4')">第 4 章</a></li>
          </ul>
        </div>
        <div data-options="region:'center',title:'Main Title',iconCls:'icon-ok'" id="center">
        </div>
    </div>
    </center>
    <script>
```

```
        function showcontent(chapter) {
            switch(chapter)
            {
              case "1":
                  $('#center').html('开发运行环境');
                  break;
              case "2":
                  $('#center').html('数据库基础');
                   break;
              case "3":
                  $('#center').html('网页基础知识');
                  break;
              case "4":
                  $('#center').html('c#基础');
                  break;
            }
         }
    </script>
</body>
</html>
```

说明：

① 代码将浏览器分成 5 个区域

north：一般存放 banner。

south：一般是版权和网页其他的声明。

west：一般用作导航。

east：可用作广告。

center：页面要表现的内容。

5 个区域中，center 是必需的，其他区域可供选择。

② 单击导航区域，执行 showcontent(chapter) 函数，采用的是 JavaScript 和 jQuery 语法。

12.3 对话框

不借助 jQuery 框架，显示对话框常用的是 JavaScript 中的 alert() 和 confirm()，显示的界面不友好。使用 jQuery EasyUI 可以显示出非常漂亮的对话框。

messager 提供了不同风格的消息框，包括 alert、confirm、prompt、progress 等，这些消息框全部是异步的，与这些对话框交互后，利用回调函数执行其他任务。

12.3.1 $.messager.show(options)

在屏幕右下角显示一个窗口。options 包括以下内容。

(1) howType：定义消息窗口显示的方式，取值有 null、slide、fade、show，默认是 slide。

(2) showSpeed：定义消息窗口显示出来需要的时间(毫秒)，默认是 600 毫秒。
(3) width：定义消息框的宽度，默认是 250。
(4) height：定义消息框的高度，默认是 100。
(5) title：定义显示在窗口标题上的文本内容。
(6) msg：窗口中显示的内容。
(7) style：消息窗口上用户自定义的显示样式。
(8) timeout：定义窗口出现后，保留的时间。设置为 0 时，只有当用户关闭时窗口才消失；设置为非 0 时，经过指定的时间后窗口自动关闭。

例如：

```
$.messager.show({
    title: 'My Title',
    msg: 'Message will be closed after 5 seconds.',
    timeout: 5000,
    showType: 'slide'
});
```

图 12-4 所示的窗口由下向上滑出，显示在屏幕右下角，5 秒后窗口自动消失。

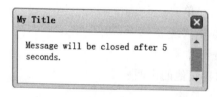

图 12-4　$.messager.show 显示的窗口

例如：

```
$.messager.show({
    title:'My Title',
    msg:'Message will be closed after 5 seconds.',
    showType:'show',
    style:{
        right:'',
        top:document.body.scrollTop + document.documentElement.scrollTop,
        bottom:''
    }
});
```

图 12-4 所示的窗口显示在屏幕上方中央的位置。

12.3.2　$.messager.alert(title,msg,icon,fn)

弹出一个 alert 窗口。各参数的含义如下。

title：窗口标题上文本。
msg：窗口中显示的内容。

icon：窗口中出现的图标，可供选择的取值有 error、question、info、warning。

fn：单击 OK 按钮后，触发的回调函数。

示例代码如下：

```
$.messager.alert('My Title','Here is a info message!','info');
$.messager.alert({
    title: 'My Title',
    msg: 'Here is a message!',
    fn: function(){
        //...
    }
});
```

12.3.3　$.messager.confirm(title,msg,fn)

带有 OK 和 Cancel 两个按钮的消息确认对话框。各参数的含义如下。

title：窗口标题上文本。

msg：窗口中显示的内容。

fn(b)：回调函数。当用户单击 OK 按钮时，将 true 传给函数；否则将 false 传给函数。

示例代码如下：

```
$.messager.confirm('Confirm', 'Are you sure to exit this system?', function(r){
    if (r){
        //exit action;
    }
});
$.messager.confirm({
    title: 'My Title',
    msg: 'Are you confirm this?',
    fn: function(r){
        if (r){
            alert('confirmed: ' + r);
        }
    }
});
```

12.3.4　$.messager.prompt(title,msg,fn)

用户能够输入文本内容，带有 OK 和 Cancel 按钮的对话框。各参数的含义如下。

title：窗口标题上文本。

msg：窗口中显示的内容。

fn(val)：带有用户输入值的回调函数。

示例代码如下：

```
$.messager.prompt('Prompt','Please enter your name:', function(r){
    if (r){
```

```
            alert('Your name is:' + r);
        }
    });
    $.messager.prompt({
        title: 'Prompt',
        msg: 'Please enter your name:',
        fn: function(r){
            if (r){
                alert('Your name is:' + r);
            }
        }
    });
```

12.3.5 $.messager.progress(options or method)

显示进度条的消息框。各参数的含义如下。

① options 定义如下。

title：窗口标题上文本。

msg：消息框中正文。

text：进度条上显示的文本。

interval：每次进度更新之间以毫秒为单位的时间长度。默认值是 300。

② method 定义如下。

bar：获得进度条对象。

close：关闭进度条窗口。

示例代码如下：

```
function messageC() {
    var win = $.messager.progress({
        title: 'Please waiting',
        msg: 'Loading data...'
    });
    setTimeout(function () {
        $.messager.progress('close');
    }, 5000)
}
```

12.4 form

12.4.1 form 提交数据

例 12-2 使用 jQuery 建立表单，完成数据提交，如图 12-5 所示。

❶ 新建立 12_2.html 文件，"源"视图中的内容如下：

```
<!DOCTYPE html>
<html xmlns = "http://www.w3.org/1999/xhtml">
```

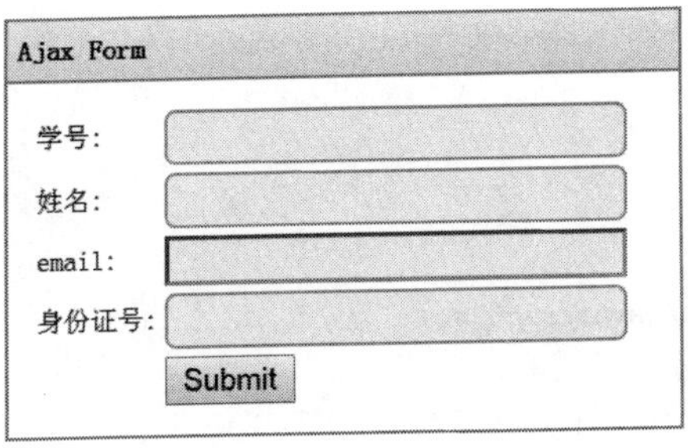

图 12-5　Ajax 表单

```html
<head>
<meta http-equiv="Content-Type" content="text/html; charset=utf-8"/>
    <title>jQuery easyUi Form</title>
    <link rel="stylesheet" type="text/css" href="../Scripts/themes/default/easyui.css"/>
    <link rel="stylesheet" type="text/css" href="../Scripts/themes/icon.css"/>
    <link rel="stylesheet" type="text/css" href="../Scripts/demo.css"/>
</head>
<body>
    <script type="text/javascript" src="../Scripts/jquery.min.js"></script>
    <script type="text/javascript" src="../Scripts/jquery.easyui.min.js"></script>
    <div class="easyui-panel" title="Ajax Form" style="width:300px;padding:10px;">
        <form id="ff" method="post">
            <table>
                <tr>
                    <td>学号:</td>
                    <td><input name="student_no" class="f1 easyui-textbox"
                        data-options="required:true"/></td>
                </tr>
                <tr>
                    <td>姓名:</td>
                    <td><input name="student_name" class="f1 easyui-textbox"
                        data-options="required:true"/></td>
                </tr>
                <tr>
                    <td>email:</td>
                    <td><input name="email" class="f1 easyui-validatebox"
                        data-options="required:true,validType:'email'"/></td>
                </tr>
                <tr>
                    <td>身份证号:</td>
                    <td><input name="identification" class="f1 easyui-textbox"
                        data-options="required:true"/></td>
                </tr>
                <tr>
                    <td></td>
                    <td><input type="submit" value="Submit"/></td>
                </tr>
            </table>
        </form>
```

```
            </div>
            <style>
                .f1{
                    width:200px;
                }
            </style>
            <script type = "text/javascript">
                $(function () {
                    $('#ff').form({
                        url: 'form1_student.ashx',
                        onSubmit: function () {
                            //检查表单,返回false时,表单不提交
                            return $(this).form('validate');
                        },
                        success: function (data) {
                            //提交成功时,将后台返回的JSON字符串转变为JSON对象
                            var jsonObj = eval("(" + data + ")")
                            $.messager.alert('Info', jsonObj.information, 'info');
                        }
                    });
                });
            </script>
    </body>
</html>
```

说明:

① <form>中没有指定action属性,表单提交是通过脚本中$('#ff').form({url:'form1_student.ashx',})指定url,单击"提交"按钮实现的。如果不使用"提交"按钮,而是普通命令按钮,如<button onclick="btnSubmit()">命令按钮提交</button>,可以通过以下脚本完成提交:

```
function btnSubmit()
{
    $('#ff').form('submit', {
      url: 'form1_student.ashx',
      onSubmit: function(){
        return $(this).form('validate');
      },
      success:function(data){
            var jsonObj = eval("(" + data + ")")
            $.messager.alert('Info', jsonObj.information, 'info');
      }
    });
}
```

② onSubmit表单事件在提交表单前触发。如果返回的是false,表单不会被提交;提交表单时可以增加其他参数,将数据一并提交到服务器:

```
$('#ff').form('submit', {
```

```
    url:...,
    onSubmit: function(param){
        param.p1 = 'value1';
        param.p2 = 'value2';
    }
});
```

③ 通过表单的 validate 方法,对表单上字段进行有效性验证。只有当表单上所有的字段都通过有效性验证时,$(this).form('validate')才返回 true。

表单中需要验证的输入值包括:所有的输入值都是必需的,通过<input data-options="required:true">指定;对输入的 email 要满足 email 的正则表达式。

④ 有效性规则通过在 data-options 中定义属性 required 和 validType 完成,validType 可以指定验证的规则有以下几个。

email:要与 email 正则表达式相匹配。

url:要与 url 正则表达式相匹配。

length[0,100]:长度介于 x 与 x 间的字符(x 是取值为 0~100 的整数)。

remote['http://.../action.do','paramName']:发送 Ajax 请求执行验证的有效性,请求成功则返回 true。

⑤ success 表单事件在表单提交成功时触发。表单提交成功后,示例中服务器程序 form1_student.ashx 返回的数据是 JSON 字符串:

{'status':'ok','information':'输入成功'}

在 success 事件的回调函数 function(data)中将 JSON 字符串转变为 JSON 对象 jsonObj。

如果服务器返回的 JSON 字符串格式不正确或者只是普通的字符串,可直接执行 $.messager.alert('Info', data, 'info');输出返回的结果。

❷ 新建一般处理程序 form1_student.ashx,代码如下:

```
public void ProcessRequest(HttpContext context)
{
    string student_no = context.Request["student_no"];
    string student_name = context.Request["student_name"];
    string email = context.Request["email"];
    string identification = context.Request["identification"];
    if (DBbase.ConnectionSqlServer() == true)              //连接数据库成功
    {
        DBbase.InsertStudent(student_no, student_name, email, identification);
        context.Response.Write("{'status':'ok','information':'输入成功'}");
        DBbase.CloseSqlServer();                            //关闭数据库
    }
}
```

说明:数据提交后,由 form1_student.ashx 处理表单的数据。将记录写入数据库的程序,直接调用 DBbase.cs 中的 InsertStudent 方法完成。如果该类库与当前文件不在同一个命名空间,需要引入 DBbase.cs 所在的命名空间,如本书中:

using aspExample.DbEntity;

12.4.2　form 加载数据

例 12-3　用表单显示 JSON 格式的数据，运行结果如图 12-6 所示。示例中 form_data1.json 的内容如下：

```
{
    "student_no":"B010101",
    "student_name":"段誉",
    "email":"abc@16.com",
    "identification":"130305201003202256"
}
```

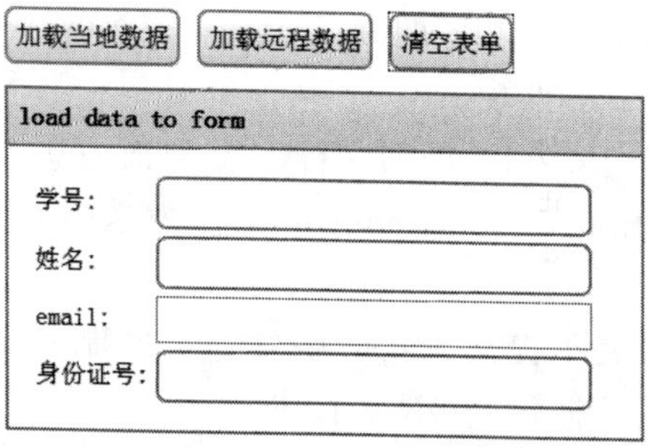

图 12-6　加载数据到表单

新建 12_3.html 文件，"源"视图中的内容如下：

```
<!DOCTYPE html>
<html xmlns = "http://www.w3.org/1999/xhtml">
<head>
<meta http-equiv = "Content-Type" content = "text/html; charset = utf-8"/>
    <title> jQuery easyUi Form </title>
    <link rel = "stylesheet" type = "text/css" href = "../Scripts/themes/default/easyui.css"/>
    <link rel = "stylesheet" type = "text/css" href = "../Scripts/themes/icon.css"/>
    <link rel = "stylesheet" type = "text/css" href = "../Scripts/demo.css"/>
</head>
<body>
    <script type = "text/javascript" src = "../Scripts/jquery.min.js"></script>
    <script type = "text/javascript" src = "../Scripts/jquery.easyui.min.js"></script>
    <div style = "margin:10px 0;">
       <a href = "javascript:void(0)" class = "easyui-linkbutton" onclick = "loadLocal()">加载当地数据</a>
       <a href = "javascript:void(0)" class = "easyui-linkbutton" onclick = "loadRemote()">加载远程数据</a>
       <a href = "javascript:void(0)" class = "easyui-linkbutton" onclick = "clearForm()">清空表单</a>
    </div>
    <div class = "easyui-panel" title = "load data to form" style = "width:300px; padding:10px;">
        <form id = "ff" method = "post">
```

```html
            <table>
                <tr><td>学号:</td>
                    <td><input name="student_no" class="f1 easyui-textbox" data-options="readonly:true"/></td>
                </tr>
                <tr>
                    <td>姓名:</td>
                    <td><input name="student_name" class="f1 easyui-textbox" data-options="readonly:true"/></td>
                </tr>
                <tr>
                    <td>email:</td>
                    <td><input name="email" class="f1 easyui-validatebox" data-options="readonly:true"/></td>
                </tr>
                <tr>
                    <td>身份证号:</td>
                    <td><input name="identification" class="f1 easyui-textbox" data-options="readonly:true"/></td>
                </tr>
            </table>
        </form>
    </div>
    <style>
        .f1{
            width:200px;
        }
    </style>
    <script type="text/javascript">
        function loadLocal(){
            $('#ff').form('load',{
                student_no:'B940105',
                student_name:'乔峰',
                email: 'abc@163.com',
                identification:'110106200010200023',
            });
        }
        function loadRemote(){
            $('#ff').form('load', 'form_data1.json');
        }
        function clearForm(){
            $('#ff').form('clear');
        }
    </script>
</body>
</html>
```

说明:表单加载数据使用方法 load,该方法可以加载本地数据,也可以加载远程数据。

本示例加载的远程数据是一个 JSON 格式的文件,如果要将查询结果加载到表单中,可以:

```
function loadRemote(){
    $('#ff').form('load', 'getSearch.ashx');
}
```

编写 getSearch.ashx,将查询结果转为 JSON 字符串即可。很显然,以 form 显示数据,只能每次显示一条记录。要显示多条记录,可使用 Datagrid。

12.5 jQuery EasyUI 的 Datagrid

使用 Datagrid 可以显示 JSON 格式的数据,可以通过 Datagrid 编辑、修改、删除数据。

例 12-4 用 Datagrid 显示出 DatagridData.json 中以下的数据,设置 unicost 大于等于 13 时,单元格设置为黄底红字,效果如图 12-7 所示。

```
{"total":4,"rows":[
    {"productid":"FI-SW-01","productname":"Koi","unitcost":10.00},
    {"productid":"K9-DL-01","productname":"Dalmation","unitcost":12.00},
    {"productid":"RP-SN-01","productname":"Rattlesnake","unitcost":15.00},
    {"productid":"RP-LI-02","productname":"Iguana","unitcost":13.00}
]}
```

product		
产品编号	产品名	单价
FI-SW-01	Koi	10
K9-DL-01	Dalmation	12
RP-SN-01	Rattlesnake	15
RP-LI-02	Iguana	13

图 12-7 jQuery EasyUI 显示数据

新建 12_4.html 文件,"源"视图中的内容如下:

```
<!DOCTYPE html>
<html xmlns="http://www.w3.org/1999/xhtml">
<head>
<meta http-equiv="Content-Type" content="text/html; charset=utf-8"/>
    <title></title>
    <link rel="stylesheet" type="text/css" href="../Scripts/themes/default/easyui.css"/>
    <link rel="stylesheet" type="text/css" href="../Scripts/themes/icon.css"/>
    <link rel="stylesheet" type="text/css" href="../Scripts/demo.css"/>
    <style>
        .main
        {
            text-align:center;
            margin:0 auto;
            width:550px;
        }
    </style>
```

```html
</head>
<body>
    <script type="text/javascript" src="../Scripts/jquery.min.js"></script>
    <script type="text/javascript" src="../Scripts/jquery.easyui.min.js"></script>
    <div class="main">
    <div style="margin:20px 0;"></div>
      <table id="tt1" class="easyui-datagrid"
           Title="product" style="width:360px;height:150px"
           data-options="url:'Datagrid.json',method:'get',singleSelect:true">
        <thead>
          <tr>
           <th field="productid" width="150">产品编号</th>
           <th field="productname" width="150">产品名</th>
           <th field="unitcost" width="50" styler="cellStyler">单价</th>
          </tr>
        </thead>
      </table>
 </div>
   <script>
       function cellStyler(value, row, index)
       {
           if (value >= 13) {
               return 'background-color:#ffee00;color:red;';
           }
       }
   </script>
</body>
</html>
```

说明：

① jQuery EasyUI 使用标准的 JSON 数据格式，包含 total 和 rows 属性。

② jQuery EasyUI 将现有的＜table＞转变为 datagrid，方法如下：

```html
<table class="easyui-datagrid">
    <thead>
        <tr>
            <th data-options="field:'code'">Code</th>
            <th data-options="field:'name'">Name</th>
            <th data-options="field:'price'">Price</th>
        </tr>
    </thead>
    <tbody>
        <tr>
            <td>001</td><td>zhang</td><td>25</td>
        </tr>
        <tr>
            <td>002</td><td>wang</td><td>45</td>
        </tr>
    </tbody>
</table>
```

③ 也可以建立＜table＞标记符后，利用＜th＞标记定义表中各行，productid、productname、unitcost 要与 JSON 中的属性名相对应。method：'get' 不可少；否则得不到数据。

④ 如果使用 JavaScript 建立 Datagrid，代码如下：

```
<table id="tt1"></table>
<script>
$('#tt1').datagrid({
    url:'Datagrid.json', method:'get',
    columns:[[
        {field:'productid',title:'产品代码',width:100},
        {field:'productname',title:'产品名称',width:100},
        {field:'unitcost',title:'单价',width:100,align:'right'}
    ]]
});
</script>
```

Datagrid 属性较多，表 12-1 罗列出一部分属性。

表 12-1　Datagrid 部分属性

属　性	值 类 型	说　　明
columns	array	Datagrid 配的列，列属性见表 12-2
frozenColumns	array	与 columns 类似，但这些列将被固定在表格的左侧
fitColumns	boolean	取 true 时列宽自动调整，以适应 grid 列宽度，防止出现水平滚动条
resizeHandle	string	取值 'left'、'right'、'both'。调整列宽的位置。只有 1.3.2 以后的版本才能用
autoRowHeight	boolean	根据行的内容，是否自动调整行高。默认为 true

toolbar	array, selector	通过 array 和 selector 两种方式，定义 Datagrid 上部工具栏。通过 <div> 选择符定义工具栏的示例：
		`$('#dg').datagrid({`
		`toolbar:'#tb'`
		`});`
		`<div id="tb">`
		`  <a href="#" class="easyui-linkbutton"`
		`      data-options="iconCls:'icon-edit',plain:true"/a>`
		`  <a href="#" class="easyui-linkbutton"`
		`      data-options="iconCls:'icon-help',plain:true"/a>`
		`</div>`

striped	boolean	行是否呈条纹显示，默认为 false
method	string	从远程请求数据的方式，默认为 post。设置 url 为 JSON 格式文件时，要设置该方法为 get 才能得到数据
nowrap	boolean	默认为 true，显示数据时不换行
idField	string	指定行的标识字段
url	string	从远程指定的 url 获取数据

续表

属性	值类型	说　　明
data	array, object	装载指定的数据，1.3.2 以后的版本可用。如： $('#dg').datagrid({ 　　data: [　　　　{f1:'value11', f2:'value12'}, 　　　　{f1:'value21', f2:'value22'} 　　] });
loadMsg	string	从远程装入数据时出现的提示信息
emptyMsg	string	无记录时出现的提示信息
pagination	boolean	默认为 false，不分页。设置为 true 时分页
rownumbers	boolean	设置为 true 时增加显示行号的列。默认设置为 false
singleSelect	boolean	设置为 true 时只能选择一行。默认设置为 false
ctrlSelect	boolean	设置为 true 时，选择行时按住 Ctrl 键，允许选择多行。默认设置为 false
selectOnCheck	boolean	设置为 true 时，单击 checkbox 时会选中该行。默认值为 true。1.3 版本以上的可用
pagePosition	string	定义分页工具栏的位置。取值 'top'、'bottom'、'both'。1.3 以上的版本可用
pageSize	number	分页时初始化每页显示的记录数
pageNumber	number	分页时初始化的页数

queryParams	object	请求远程数据时，发送的请求符合数据的条件，如：
		$('#dg').datagrid({
		　　queryParams: {
		　　　　name: 'easyui',
		　　　　subject: 'datagrid'
		　　}
		});

rowStyler	function	返回行显示的样式，如 'background:red'。函数具有 rowIndex 和 rowData 两个参数。rowIndex 是从 0 开始的行号；rowData 是响应的当前行的记录。示例：
		$('#dg').datagrid({
		　rowStyler: function(index,row){
		　　if (row.listprice>80){
		　　　return 'background-color:#6293BB;color:#fff;';
		　　}
		　}
		});

loader	function	定义如何从远程装载数据。返回 false 时可取消装载，有 3 个参数。 param：传递到远程的参数对象 success(data)：成功接收数据后的回调函数 error()：接收数据失败时的回调函数

续表

属性	值类型	说明
loadFilter	function	返回要显示的过滤后的数据。函数的参数 data 和 and 'rows'属性的标准的数据对象
editors	object	定义编辑行时行中出现的编辑对象

columns 具有的属性见表 12-2。

表 12-2 columns 主要属性

属性	值类型	说明
title	string	列标题
field	string	列对应的字段名
width	number	列宽，不设置，根据内容自动调节。但会降低性能
rowspan	number	合并 number 行
colspan	number	合并 number 列
align	string	列数据对齐方式，取值包括'left'、'right'、'center'
halign	string	列标题对齐方式，取值包括'left'、'right'、'center'，不设置，与 align 相同
sortable	boolean	列是否排序，设置为 true 时排序
resizable	boolean	列宽度是否允许调整，取值为 true、false
hidden	boolean	设置为 true 时隐藏列
checkbox	boolean	是否显示 checkbox
formatter	function	单元格格式函数。函数有 3 个参数。 value：字段值 rowData：行记录的数据 rowIndex：行编号

styler	function	单元格显示风格。函数有 3 个参数。
		value：字段值
		rowData：行记录的数据
		rowIndex：行编号。如果要将 unitcost >= 15 的单元格前景色设置为红色，背景设置为黄色，代码为：
		{
		    field: 'unitcost', title: '单价', width: 100,
		    styler: function (value, row, index) {
		        if (value >= 15) {
		            return 'background-color:#ffee00;color:red;';
		    }
		}

editor	string, object	编辑的类型，object 参数可选。string 指定的类型，包括 text、textbox、numberbox、numberspinner、combobox、combotree、combogrid、datebox、datetimebox、timespinner、datetimespinner、textarea、checkbox、validatebox

例 12-5 用 jQuery EasyUI 的 Datagrid 显示 student 表中的全部记录,效果如图 12-8 所示。

学生全部记录				
学号	姓名	性别	email	身份证号
30012035	楚云飞	男	zsf@sohu.com	230304200611201537
40012030	赵三	男	zs@sohu.com	230304200010302378
41321059	李孝诚	男	lxc@163.com	110106199802030013
41340136	马小玉	女	mxy@126.com	130304199705113028
41355062	王长林	男	wcl@163.com	110106199512035012
41361045	李将寿	男	ljc@126.com	145106199012035013X
41361258	刘登山	男	lds@126.com	130030520001234235
41361260	张文丽	女	zhangwl@163.com	140305200106202348
41401007	鲁宇星	男	lyx@sohu.com	110108200012035012
41405002	王小月	女	wxy@263.com	130304199901115047
41405003	张晨露	女	zcl@263.com	110107199810038022
41405005	张伟达	男	zwd@263.net	120305199010270018
41405221	张康林	女	zkl@126.net	130305199805231248
41405238	张雨	女	zy@sohu.com	230304200210052346

图 12-8 Datagrid 显示 student 表中全部记录

❶ 新建页面 12_5.html,视图"源"中的内容如下:

```
<!DOCTYPE html>
<html xmlns="http://www.w3.org/1999/xhtml">
<head>
<meta http-equiv="Content-Type" content="text/html; charset=utf-8"/>
<title></title>
<link rel="stylesheet" type="text/css" href="../Scripts/themes/default/easyui.css"/>
<link rel="stylesheet" type="text/css" href="../Scripts/themes/icon.css"/>
<link rel="stylesheet" type="text/css" href="../Scripts/demo.css"/>
<style>
    .main
    {
        text-align:center;
        margin:0 auto;
        width:550px;
    }
</style>
</head>
<body>
    <script type="text/javascript" src="../Scripts/jquery.min.js"></script>
    <script type="text/javascript" src="../Scripts/jquery.easyui.min.js"></script>
    <div class="main">
        <h2 style="text-align:center">jQuery EasyUI 中 Datagrid 的基本用法</h2>
        <div style="margin:20px 0;"></div>
        <table id="tt1" class="easyui-datagrid" title="学生全部记录" style="width:550px;height:400px" data-options="url:'operateStudent.ashx?op=1'">
            <thead>
```

```html
            <tr>
                <th field = "student_no" width = "100">学号</th>
                <th field = "student_name" width = "100">姓名</th>
                <th field = "gender" width = "50">性别</th>
                <th field = "email" width = "100">email</th>
                <th field = "identification" width = "180">身份证号</th>
            </tr>
        </thead>
    </table>
</div>
</body>
</html>
```

说明:

① 在<table>中通过设置 class="easyui-datagrid"将表格转变为 Datagrid,在 data-options 属性中,通过指定 URL 设置 Datagrid 的数据源为 getAllStudent.ashx。数据源可以直接指定为某个 JSON 格式文件,也可以是能够生成 JSON 的程序。

② <th field="student_no" width="100">学号</th>中,field 属性的设置要与 getAllStudent.ashx 输出的 JSON 字符串中的属性相对应。

③ data-options=url:'operateStudent.ashx? op=1',指定 Datagrid 的数据源的 URL,传递变量 op=1 到 operateStudent.ashx 文件中,根据不同的参数执行不同的动作。

❷ 新建 operateStudent.ashx 一般处理程序:

```csharp
using System.Data;
using System.Text;
using System.Data.SqlClient;
public void ProcessRequest(HttpContext context)
public class operateStudent:IHttpHandler
{

    public void ProcessRequest(HttpContext context)
    {
        string student_no, student_name, gender, email, identification, id;
        context.Response.ContentType = "text/plain";
        //操作类型,包括查询、删除、更新、插入等
        string op = context.Request["op"];/
        student_no = context.Request["student_no"];
        student_name = context.Request["student_name"];
        gender = context.Request["gender"];
        email = context.Request["email"];
        identification = context.Request["identification"];
        id = context.Request["id"];
        switch (op)
        {
            case "1"://查询
                SearchStudent(context);
                break;
            case "2"://插入
                AddStudent(context,student_no,student_name,email,identification);
```

```csharp
                break;
            case "3"://删除
                DeleteStudent(context);
                break;
            case "4"://更新
                UpdateStudent(context, student_no, student_name, email,
                    identification, Convert.ToInt16(id));
                break;
        }
    }
    public void SearchStudent(HttpContext context)
    {
        DataTable dt = new DataTable();
        DataSet ds = new DataSet();
        CommandQuery(ds, "select * from student");
        dt = ds.Tables[0];
        context.Response.Write(DBbase.DataTable2Json(dt));
    }
}
```

说明：

① 本示例只显示 student 表中的数据，只用到 op=1 的情况。

② 第 9 章例 9-6 中服务器上使用 Newtonsoft.Json 将数据表转换为 JSON，然后在 Datagrid 中显示。下面介绍另一种利用 StringBuilder 类字符串拼接的方法，拼接出满足 Datagrid 数据要求的 JSON 字符串，将其编写为 DBbase.cs 类中的一个函数，代码如下：

```csharp
///<summary>
///将数据表转换为 Datagrid 需要的 JSON
///</summary>
///<param name="dt">DataTable</param>
///<returns></returns>
public static string DataTable2Json(DataTable dt)
{
    //由表转换为 json
    StringBuilder jsonBuilder = new StringBuilder();
    jsonBuilder.Append("{\"total\"" + ":" + dt.Rows.Count + ",\"rows\":[");
    for (int i = 0; i < dt.Rows.Count; i++)
    {
        jsonBuilder.Append("{");
        for (int j = 0; j < dt.Columns.Count; j++)
        {
            jsonBuilder.Append("\"");
            jsonBuilder.Append(dt.Columns[j].ColumnName);
            jsonBuilder.Append("\":\"");
            jsonBuilder.Append(dt.Rows[i][j].ToString().Trim());
            jsonBuilder.Append("\",");
        }
        if (dt.Columns.Count > 0)
        {
            jsonBuilder.Remove(jsonBuilder.Length - 1, 1);
```

```
                }
                jsonBuilder.Append("},");
        }
        if (dt.Rows.Count > 0)
        {
            jsonBuilder.Remove(jsonBuilder.Length - 1, 1);
        }
        jsonBuilder.Append("]}");
        return jsonBuilder.ToString();
}
```

可以看出，使用 StringBuilder 可以拼接出任何格式的 JSON 字符串，但书写比较麻烦，容易出错。

例 12-6 用 jQuery EasyUI 中的 Datagrid 编辑、增加、删除 student 中的数据，运行效果如图 12-9 所示。

图 12-9 Datagrid 编辑数据

❶ 新建文件 12_6.html，"源"视图下的内容如下：

```
<!DOCTYPE html>
<html xmlns="http://www.w3.org/1999/xhtml">
<head>
<meta http-equiv="Content-Type" content="text/html;charset=utf-8"/>
    <title></title>
    <link rel="stylesheet" type="text/css" href="../Scripts/themes/default/easyui.css"/>
    <link rel="stylesheet" type="text/css" href="../Scripts/themes/icon.css"/>
    <link rel="stylesheet" type="text/css" href="../Scripts/demo.css"/>
    <style>
        .main{text-align:center; margin:0 auto;width:550px;}
    </style>
</head>
```

```html
<body>
    <script type="text/javascript" src="../Scripts/jquery.min.js"></script>
    <script type="text/javascript" src="../Scripts/jquery.easyui.min.js"></script>
    <div class="main">
    <div style="margin:10px 0"></div>
        <table id="tt" class="easyui-datagrid"
            title="学生全部记录" style="width:560px;height:400px"
            data-options="
            url:'operateStudent.ashx?op=1',
            iconCls:'icon-edit',
            striped:true,
            toolbar:'#tb',
            method:'get',
            onClickCell: onClickCell,
            onEndEdit: onEndEdit">
        <thead>
            <tr>
            <th data-options="field:'ck',checkbox:true"></th>
            <th data-options="field:'student_no',width:100,editor:'textbox'">学号</th>
            <th data-options="field:'student_name',width:100,editor:'textbox'">姓名</th>
            <th data-options="field:'gender',width:50">性别</th>
            <th data-options="field:'email',width:100,editor:'textbox'">email</th>
            <th data-options="field:'identification',width:180,editor:'textbox'">身份证号</th>
            <th data-options="field:'id',width:180,hidden:true">id</th>
            </tr>
        </thead>
        </table>
//定义工具栏
    <div id="tb" style="height:auto">
        <a href="javascript:void(0)" class="easyui-linkbutton"
        data-options="iconCls:'icon-add',plain:true" onclick="append()">Append</a>
        <a href="javascript:void(0)" class="easyui-linkbutton"
        data-options="iconCls:'icon-remove',plain:true"
        onclick="removeit()">Remove</a>
        <a href="javascript:void(0)" class="easyui-linkbutton"
        data-options="iconCls:'icon-save',plain:true"
        onclick="accept()">Accept</a>
        <a href="javascript:void(0)" class="easyui-linkbutton"
        data-options="iconCls:'icon-undo',plain:true"
        onclick="reject()">Reject</a>
        <a href="javascript:void(0)" class="easyui-linkbutton"
        data-options="iconCls:'icon-search',plain:true"
        onclick="getChanges()">Update</a>
    </div>
    </div>
    <script>
        var editIndex = undefined;
        //结束行编辑时
        function endEditing() {
            if (editIndex == undefined) {return true}
```

```javascript
        if ($('#tt').datagrid('validateRow', editIndex)) {
            $('#tt').datagrid('endEdit', editIndex);
            editIndex = undefined;
            return true;
        } else {
            return false;
        }
    }
    //单击单元格时
    function onClickCell(index, field) {
        if (editIndex != index) {
            if (endEditing()) {
                $('#tt').datagrid('selectRow', index)
                    .datagrid('beginEdit', index);
                var ed = $('#tt').datagrid('getEditor', { index: index, field: field });
                if (ed) {
                    ($(ed.target).data('textbox') ? $(ed.target).textbox('textbox') : $(ed.target)).focus();
                }
                editIndex = index;
            } else {
                setTimeout(function () {
                    $('#tt').datagrid('selectRow', editIndex);
                }, 0);
            }
        }
    }
    //结束编辑
    function onEndEdit(index, row) {
        var ed = $(this).datagrid('getEditor', {
            index: index,
            field: 'gender'
        });
        row.gender = $(ed.target).combobox('getText');
    }
    //单击工具栏上的 append 按钮,表中增加一行
    function append() {
        if (endEditing()) {
            $('#tt').datagrid('appendRow', { gender: '男' });
            editIndex = $('#dg').datagrid('getRows').length - 1;
            $('#tt').datagrid('selectRow', editIndex)
                    .datagrid('beginEdit', editIndex);
        }
    }
    //单击工具栏上的 remove 按钮,从数据库中删除选定记录
    function removeit() {
        var ids = [];
        //得到所有数据行
        var rows = $('#tt').datagrid('getSelections');
        for (var i = 0; i < rows.length; i++) {
            //向数组 ids 的末尾添加一个或者多个元素,并返回数组新的长度
```

```javascript
                ids.push(rows[i].id);
            }
            //数组中所有元素组成一个字符串 deleteID
            var deleteID = ids.join(',');
            if (deleteID.length > 0) {
                $.messager.confirm('Confirm', 'Are you sure you want to destroy this record?',
function (r) {
                    if (r) {
                        $.post('operateStudent.ashx?op=3', {id: deleteID}, function (result) {
                            if (result == "ok") {
                                $('#tt').datagrid('reload');    //重新加载数据
                                $.messager.alert('information', 'delete finished!');
                            } else {
                                $.messager.show({                //显示出错信息
                                    title: 'Error',
                                    msg: result.errorMsg
                                });
                            }
                        });
                    }
                });
            }
        }
        //单击工具栏上的 accept 按钮,接收数据修改,但还没有存到数据库
        function accept() {
            if (endEditing()) {
                $('#tt').datagrid('acceptChanges');
            }
        }
        //单击工具栏上的 reject 按钮
        function reject() {
            $('#tt').datagrid('rejectChanges');
            editIndex = undefined;
        }
        //单击工具栏上的 getChanges 按钮,批量更新数据库
        function getChanges() {
            var rows = $('#tt').datagrid('getChanges');
            //批量更新到数据库中
            for (var i = 0; i < rows.length; i++)
            {
                $.post('operateStudent.ashx?op=4', { id: rows[i].id, student_no: rows[i].student_no, student_name: rows[i].student_name, email: rows[0].email, identification: rows[i].identification }, function (result) {
                    if (result == "ok") {
                        $('#tt').datagrid('reload');
                        $.messager.alert('information', 'Update finished!');
                    }
                    else {
                        $.messager.alert('information', 'Update failed!');
                        reject();
                    }
```

```
                });
            }
        }
    </script>
</body>
</html>
```

说明：

① Datagrid 的事件 onClickCell，当用户单击单元格时触发。具有参数 index、field、value。index 表示行编号，field 表示字段名，value 表示当前记录的值。示例中 onClickCell：onClickCell 表示单击单元格执行 onClickCell 函数。

② <th data-options="field:'student_no', width:100, editor:'textbox'">学号</th>，编辑"student_no"字段时使用的 editor 是 textbox。

③ $('#tt').datagrid('selectRow', index)选中 index 行。

$('#tt').datagrid('beginEdit', index)编辑 index 行。

④ var ed = $('#tt').datagrid('getEditor', { index: index, field: field })得到 index 行 field 字段指定的 editor。GetEditor 有两个参数：index 表示行索引；field 表示字段名。例如：

```
//get the datebox editor and change its value
var ed=$('#dg').datagrid('getEditor', {index:1,field:'birthday'});
$(ed.target).datebox('setValue', '5/20/2014');
```

其中，$(ed.target)表示选定的单元格。

⑤ 函数 append()中 $('#tt').datagrid('appendRow', { gender:'男' })；增加一行，设置 gender 的值为'男'，示例没有提供将数据保存到数据库的代码。

⑥ 函数 accept()中 $('#tt').datagrid('acceptChanges')；提交数据装载后或者上次修改后的所有改动。

⑦ 函数 reject()中 $('#tt').datagrid('rejectChanges')；放弃数据装载后或者上次修改后的所有改动。

⑧ 函数 removeit()批量删除数据。$.messager.confirm()是 jQuery EasyUI 提供的对话框，单击 OK 按钮将删除记录。

```
$.post('operateStudent.ashx?op=3', { id: deleteID }, function (result) {
})
```

向服务器程序 operateStudent.ashx 发送信息，想要发送给服务器的数据有 op=3 和 id 为变量 deleteID 的值，也可以写为：

```
$.post('operateStudent.ashx?op=3&id=' + deleteID, function (result) {
})
```

或者

```
$.post('operateStudent.ashx',{op:3,id:deleteID }, function (result) {
})
```

function(result)是回调函数,result 是后台程序 operateStudent.ashx 中 DeleteStudent()函数中 context.Response.Write("ok")输出的值,即 OK。

⑨ function getChanges()中 $.post()的用法与上面相同。

❷ 在例 12-5 的 operateStudent.ashx 后台程序文件中添加以下代码:

```
//保存数据到数据库
public void AddStudent(HttpContext context, string student_no, string student_name, string email, string identification)
{
    DBbase.InsertStudent(student_no,student_name,email,identification);
    context.Response.Write("ok");                    //输出操作完成的提示 OK
}
//更新数据到数据库
public void UpdateStudent(HttpContext context,string student_no, string student_name, string email, string identification, int id)
{
    DBbase.UpdateStudent(id, student_no, student_name, email, identification);
//更新操作完成的提示"ok",在 12-4.html 中通过 getChanges()的回调函数 function(result)获得
输出,放入 result 中
    context.Response.Write("ok");
}
//删除数据库中指定的数据
public void DeleteStudent(HttpContext context)
{
    string id = context.Request["id"];
    DBbase.CommandNonQuery("delete from student where id in(" + id + ")");
    context.Response.Write("ok");                    //输出操作完成的提示 OK
}
```

❸ 运行 12_6.html。

上面示例中用到了 Datagrid 的一些方法,表 12-3 给出了这些方法的解释和部分示例。

表 12-3 Datagrid 的方法

方 法	说 明

load	装载并显示第 1 页的数据行,用法:
	`$('#dg').datagrid('load',{`
	`    code: '01',`
	`    name: 'name01'`
	`});`

reload	重新装载数据,与 load 相同但停留在当前页而非第 1 页
fitColumns	列自动缩放适应 datagrid 的宽度
fixColumnSize	固定列的大小,例如: `$('#dg').datagrid('fixColumnSize', 'name');` //固定 'name'列宽 `$('#dg').datagrid('fixColumnSize');` //固定所有的列宽
loadData	装载数据,原来的数据从 Datagrid 中移除
getData	返回已装载的数据
getRows	返回当前页的行数据

续表

方　　法	说　　明
getSelected	返回第一个选中的行记录。若没有行被选中，则返回 null
getSelections	返回所有选中的行。若没有行被选中，则返回 null 数组
clearSelections	清除所有选择
selectAll	选中当前页的所有行
selectRow	选中指定的行，如 $('#dg').datagrid('selectRow', 3);，选中第 4 行
beginEdit	编辑指定的行，如 $('#dg').datagrid('beginEdit', 3);，编辑第 4 行
endEdit	结束指定行的编辑，如 $('#dg').datagrid('endEdit', 3);，结束第 4 行的编辑
cancelEdit	取消指定行的编辑
getEditor	获得指定的 editor。方法有两个参数。index：行号；field：字段名。如获得 datebox 的 editor 并改变其值： var ed = $('#dg').datagrid('getEditor', {index:1,field:'birthday'}); $(ed.target).datebox('setValue', '5/4/2012');

updateRow	更新指定的行。方法有两个参数。index：指定行号；row：新的行数据。例如
	$('#dg').datagrid('updateRow',{
	　　index: 2,
	　　row: {
	　　　　name: 'new name',
	　　　　note: 'new note message'
	　　}
	});

appendRow	增加一新行，新行被加到表的最后。例如： $('#dg').datagrid('appendRow',{ 　　name: 'new name', 　　age: 30, 　　note: 'some messages' });

insertRow	插入新行。方法有两参数。index：插入到的行编号；row：插入的行数据。例如：
	//i 在第 2 行插入数据
	$('#dg').datagrid('insertRow',{
	　　index: 1,　　　//行编号从 0 开始
	　　row: {
	　　　　name: 'new name',
	　　　　age: 30,
	　　　　note: 'some messages'
	　　}
	});

deleteRow	删除指定的行
getChanges	获得上次数据提交后数据发生改变的行。可以使用参数指定发生改变的类型，包括 inserted、deleted、updated 等。如果不指定参数，指所有行的改变
acceptChanges	提交装载后或者最后一次执行完 acceptChanges 后发生改变的数据
rejectChanges	回滚装载后或者最后一次执行完 acceptChanges 后发生改变的数据

方　　法	说　　明
showColumn	显示指定的字段
hideColumn	隐藏指定的字段

sort	为 Datagrid 排序,1.3.6 版本以上的可用。例如:
	//根据'itemid'排序
	$('#dg').datagrid('sort', 'itemid');
	//按指定的列'productid'以降序排序
	$('#dg').datagrid('sort', {
	　　sortName: 'productid',
	　　sortOrder: 'desc'
	});
	//按先'productid'列降序,再'listprice'列升序的方式排序
	$('#dg').datagrid('sort', {
	　　sortName: 'productid,listprice',
	　　sortOrder: 'desc,asc'
	});

习题与思考

(1) order.json 中的内容如下:

```
[
    {"dateTime":"2018-4-3","status":"已下单","content":"订单等待配货"},
    {"dateTime":"2018-4-5","status":"已发货","content":"卖家发货"},
    {"dateTime":"2018-4-5","status":"已揽件","content":"上海松江公司收货员:丁顺华"}
]
```

用 jQuery EasyUI 的 Datagrid 显示出 order.json 中的数据,如图 12-10 所示。

时间	订单状态	详情
2018-4-3	已下单	订单等待配货
2018-4-5	已发货	卖家发货
2018-4-5	已揽件	上海松江公司收货员:丁顺华

图 12-10　运行结果

(2) 将课程表中的记录用 Datagrid 显示出来,效果如图 12-11 所示。选中一条记录,单击 GetSelected 链接,效果如图 12-12 所示;选中多条记录,单击 GetSelections 链接,效果如图 12-13 所示。要求图 12-12 和图 12-13 弹出来的窗口在 5 秒后自动消失。

(3) 将 student 表中记录列出来,如图 12-14 所示,选中记录后,单击"删除记录"按钮,弹出是否删除对话框,如图 12-15 和图 12-16 所示。

(4) form_data2.json 文件中的内容如下:

{"产品编号":"FI-SW-01","单价":10.00,"数量":500}

将该文件中的内容用 form 显示出来,运行效果如图 12-17 所示。

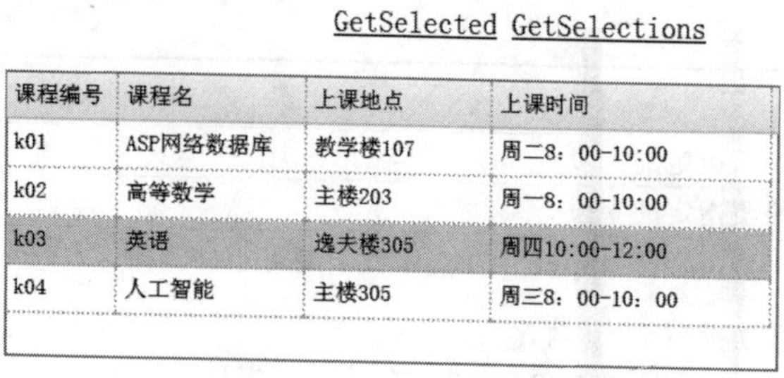

课程编号	课程名	上课地点	上课时间
k01	ASP网络数据库	教学楼107	周二8：00-10:00
k02	高等数学	主楼203	周一8：00-10:00
k03	英语	逸夫楼305	周四10:00-12:00
k04	人工智能	主楼305	周三8：00-10：00

图 12-11　练习运行结果

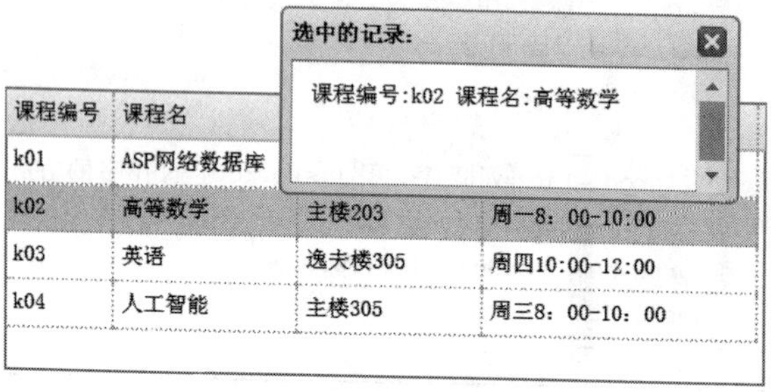

课程编号	课程名		
k01	ASP网络数据库		
k02	高等数学	主楼203	周一8：00-10:00
k03	英语	逸夫楼305	周四10:00-12:00
k04	人工智能	主楼305	周三8：00-10：00

图 12-12　选中一条记录并单击 GetSelected 的运行结果

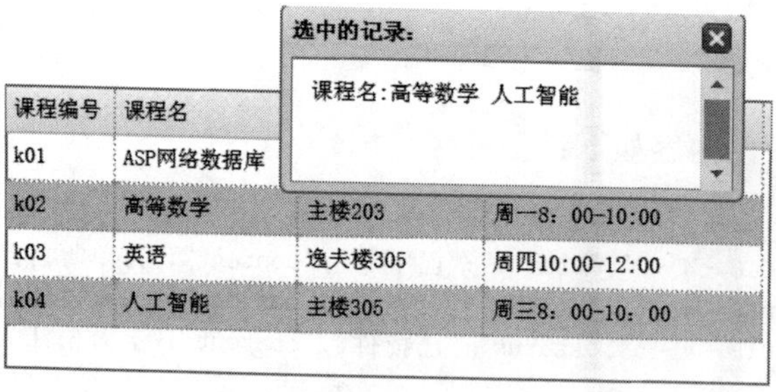

课程编号	课程名		
k01	ASP网络数据库		
k02	高等数学	主楼203	周一8：00-10:00
k03	英语	逸夫楼305	周四10:00-12:00
k04	人工智能	主楼305	周三8：00-10：00

图 12-13　选中多条记录并单击 GetSelections 的运行结果

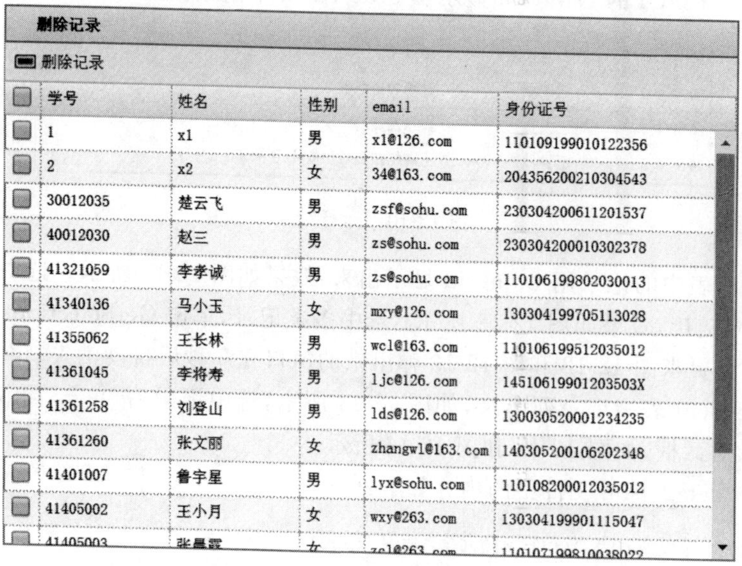

学号	姓名	性别	email	身份证号
1	x1	男	x1@126.com	110109199010122356
2	x2	女	34@163.com	204356200210304543
30012035	楚云飞	男	zsf@sohu.com	230304200611201537
40012030	赵三	男	zs@sohu.com	230304200010302378
41321059	李孝诚	男	zs@sohu.com	110106199802030013
41340136	马小玉	女	mxy@126.com	130304199705113028
41355062	王长林	男	wcl@163.com	110106199512035012
41361045	李将寿	男	ljc@126.com	14510619901203503X
41361258	刘登山	男	lds@126.com	130030520001234235
41361260	张文丽	女	zhangwl@163.com	140305200106202348
41401007	鲁宇星	男	lyx@sohu.com	110108200012035012
41405002	王小月	女	wxy@263.com	130304199901115047
41405003	张晨霞	女	zcl@263.com	110107199810038022

图 12-14　显示 student 中记录

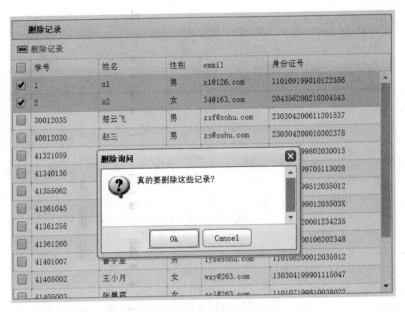

图 12-15 删除确认对话框

图 12-16 删除完成提示框

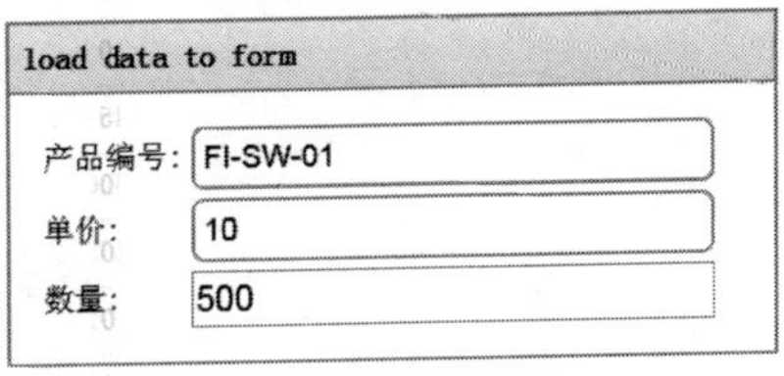

图 12-17 用 form 列出 form_data2.json 中的内容

参 考 文 献

[1] 段克奇. ASP.NET 基础教程[M]. 2 版. 北京：清华大学出版社,2014.
[2] Jonathan Chaffer,Karl Swedberg. jQuery 基础教程[M]. 4 版. 李松峰,译. 北京：人民邮电出版社,2013.
[3] Ryan Benedetti,Ronan Cranley. Head First jQuery(中文版)[M]. 林琪,译. 北京：中国电力出版社,2014.
[4] Matthew MacDonald,Adam Freeman,Mario Szpuszta. ASP.NET 4 高级程序设计[M]. 博思工作室,译. 北京：人民邮电出版社,2011.
[5] Andrew Troelsen. 精通 C#[M]. 6 版. 姚琪琳,译. 北京：人民邮电出版社,2017.
[6] 张正礼,陈作聪. ASP.NET 从入门到精通[M]. 北京：清华大学出版社,2015.
[7] 李春葆,蒋林,喻丹丹,等. ASP.NET 4.5 动态网站设计教程[M]. 北京：清华大学出版社,2016.
[8] 徐会杰,朱海,王凤科. ASP.NET 4.5 程序设计基础教程(C#版)[M]. 北京：中国工信出版集团,2016.
[9] Anderson Stellman,Jennifer Greene. Head First C#(中文版)[M]. 林琪,译. 北京：中国电力出版社,2016.
[10] Elisabeth Robson,Eric Freeman. Head First HTML 与 CSS[M]. 林琪,译. 北京：中国电力出版社,2013.
[11] Karli Watson,Jacob Vibe Hammer,Jon D Reid,等. C#入门经典. (C# 6.0 & Visual Studio 2015)[M]. 7 版. 乔立波,译. 北京：清华大学出版社,2016.
[12] Itzik Ben-Gan. SQL Server 2012 T-SQL 基础教程[M]. 张洪举,译. 北京：人民邮电出版社,2013.
[13] https://www.jstree.com,介绍 jsTree 插件.
[14] http://www.jeasyui.com/index.php,介绍 jQuery EasyUI.
[15] http://www.runoob.com,介绍 HTML、CSS、JavaScript 等.
[16] http://www.w3school.com.cn/html/index.asp,介绍 HTML、CSS、JavaScript 等.
[17] https://github.com/dabeng/OrgChart,介绍组织插件 OrgChart.
[18] https://blog.csdn.net/u013777676/article/details/53107699,介绍其他的组织插件.
[19] http://plugins.jquery.com/cookie,介绍 jquery.cookie.js 插件.
[20] https://www.npmjs.com/package/jquery-mockjax,介绍 mockjax 插件.